Improving Your NCAA® Bracket with Statistics

ASA-CRC Series on
Statistical Reasoning in Science and Society

Series Editors

Nicholas Fisher
University of Sydney, Australia

Nicholas Horton
Amherst College, MA

Deborah Nolan
University of California, Berkeley, CA

Regina Nuzzo
Gallaudet University, Washington, DC

David J Spiegelhalter
University of Cambridge, UK

Published Titles

Errors, Blunders, and Lies
How to Tell the Difference
David S. Salsburg

Visualizing Baseball
Jim Albert

Data Visualization
Charts, Maps and Interactive Graphics
Robert Grant

Improving Your NCAA® Bracket with Statistics
Tom Adams

For more information about this series, please visit:
https://www.crcpress.com/go/asacrc

Improving Your NCAA® Bracket with Statistics

Tom Adams

CRC Press
Taylor & Francis Group
Boca Raton London New York

CRC Press is an imprint of the
Taylor & Francis Group, an **informa** business
A CHAPMAN & HALL BOOK

CRC Press
Taylor & Francis Group
6000 Broken Sound Parkway NW, Suite 300
Boca Raton, FL 33487-2742

© 2019 by Taylor & Francis Group, LLC
CRC Press is an imprint of Taylor & Francis Group, an Informa business

No claim to original U.S. Government works

Printed on acid-free paper

International Standard Book Number-13: 978-1-138-59774-7 (Paperback)
978-1-138-59778-5 (Hardback)

Library of Congress Cataloging-in-Publication Data

Names: Adams, Tom.
Title: Improving your NCAA bracket with statistics / Tom Adams.
Description: Boca Raton, Florida : CRC Press, 2018. | Series: ASA-CRC series on statistical reasoning in science and society | Includes bibliographical references and index.
Identifiers: LCCN 2018045347| ISBN 9781138597785 (hardback : alk. paper) | ISBN 9781138597747 (pbk. : alk. paper) | ISBN 9780429486760 (e-book)
Subjects: LCSH: NCAA Basketball Tournament. | Basketball--Statistics. | Basketball--Betting--Statistical methods.
Classification: LCC GV885.49.N37 A34 2018 | DDC 796.323/63--dc23
LC record available at https://lccn.loc.gov/2018045347

Visit the Taylor & Francis Web site at
http://www.taylorandfrancis.com

and the CRC Press Web site at
http://www.crcpress.com

MIX
Paper from
responsible sources
FSC® C013985

Printed in the United Kingdom
by Henry Ling Limited

Contents

Preface

M OST BRACKET POOL PLAYERS will be surprised to discover that there is a body of knowledge that will greatly improve their ability to compete in a bracket pool. This knowledge has not been easy to discover. An Internet search will not reliably locate it because there is so much bad bracket advice on the Internet that competes with the best sources. You are unlikely to learn it around the office watercooler. The sharpies in your office, if there are any, are your competition in your office pool and are also unlikely to help you improve your bracket.

This book seeks to be a comprehensive guide to this knowledge and the grounds for believing it. Using charts and pictures, and avoiding complex equations, it takes you on a tour of the over 20 years of academic research on bracket improvement methods.

Brief bracket improvement tips that arise from the research are prominently displayed throughout for those more inclined to just cut to the chase. But, be warned that some of the tips may be counterintuitive, so you may need a bit of convincing by looking at the reasoning that give rise to the tips. Also, Chapter 10 seeks to provide a comprehensive presentation of the available good advice sources for bracket improvement.

The book provides a guide to some foundational ideas and methods in game theory, economics, psychology, probability, statistics, and machine learning. Basic principles are explained and historical sketches of some of the originators are included.

The basics of the bracket pool itself are explained so that even the reader unfamiliar with the pool will not be in the dark.

Bracket improvement is the book's hook, but it is not really my goal in writing the book. My goal is to inspire you with my passion for discovery in mathematics and statistics. I'm wagering that I will succeed.

Tom Adams
Cary, North Carolina

Acknowledgments

I am grateful to Lara Spieker and Rob Calver for guiding me through the logistics of book preparation and arranging some fine reviewers for early drafts. I want to thank Brad Carlin, Ed Kaplan, Bryan Clair, and Jarad Niemi for almost two decades of helpful email conversations about improving brackets. I want to thank my wife Nancy for serving as my first reviewer and tolerating my long hours in the writing room.

CHAPTER 1

The Birth of the Pool

IT ALL BEGAN AT Jody's Club Forest. Jody's bills itself as "A Taste of Irish in Staten Island." The family-friendly North Shore West Brighton neighborhood eatery and bar organized the first known bracket pool in 1977. The entry fee was $10 and 88 patrons filled out brackets (Rushin 2009). The bracket had only 32 teams that year. This first bracket pool involved picking only the Final Four®, the champion, and the number of points scored in the championship game, not the full bracket. Marquette beat North Carolina in the championship game that year.

A year later, Bob Stinson, a U.S. Postal Service employee, started a bracket pool that may stand as the first using the typical modern format and rules in Louisville, Kentucky. According to Stinson, before that people would just pick random team names out of a hat and the newspaper did not print a bracket. He wanted a contest that rewarded knowledge of the game of basketball. The pool had 15 participants in its first year (Hill 1997). Tim Trowbridge of Kent, Ohio independently started a pool that involved picking a winner for every game in the bracket back in 1981 (Allard 2017). The NCAA tournament had 48 teams back then. The National Collegiate Basketball Association, or NCAA, is the governing body for much of U.S. college basketball. Trowbridge and his friend Jeff Hunt managed the pool. They worked out the format over a few beers in a local pub. Their goal was to "once and for all determine who knew the most about college basketball." This pool also had 15 entries the first year and that grew to 200 entries by the fourth or fifth year. The filled-out bracket entry sheets were collected from local bars. An engraved plaque memorializing the names of all the winners of the "Trowbridge Hunt & Trowbridge Annual NCAA Tournament" used to hang in the Rusty Nail bar. It now hangs near the door of Trowbridge's office. The first description of an "office pool" involving the NCAA tournament bracket was in 1984 among books and sources indexed by Google (Wayne 1984).

1.1 THE TOURNAMENT

In the first intercollegiate basketball game, played in 1896, the Chicago Maroons beat the Iowa Hawkeyes. Yale was a powerhouse team in the early years. The Ivy League was the first conference. A conference is an organized group of teams in a region that mostly play each other. The Southern Conference held the first college tournament at the end of its regular season in 1921 (ESPN 2009).

Tournaments caught on. Tournaments are interesting and exciting for a number of reasons. These are single elimination tournaments—so each game is a sudden death match that (in the case of the NCAA tournament) ends the season for the losing team. Since the seniors on the team will graduate or lose eligibility, it marks the last game that a losing team would ever play together. This can be a very emotional moment. Only one team emerges from the tournament victorious in all games. Another appeal of a tournament is that it tends to be an equalizer. The best team in a conference may be favored to win all the conference games it plays all season. But playing in a tournament is like running a gauntlet. It's often the case that even the best team is more likely than not to lose the tournament. A tournament is a challenge for even the best teams, and it gives every team a chance to shock the rest and emerge victorious. This level of unpredictability is riveting for the fans.

The first NCAA tournament, consisting of eight teams, was held in 1939. It expanded to 16 teams in 1951. In 1953, the NCAA tournament added six more teams; this was the first unbalanced bracket requiring six play-in games. The tournament expanded to 32 teams in 1975, and 64 teams in 1985. In 2011, the field expanded to 68 teams, requiring four play-in games, but most bracket pools ignore the play-in games and use a balanced format involving the 64 teams that include the winners of the play-in games. (The NCAA has started calling the play-in round of four games the "first round." But this book will stick to the earlier convention of calling the round of 64 teams the "first round.")

1.2 THE BRACKET POOL EMERGES

A number of events happened around 1977 that might explain why the bracket pool caught on. UCLA dominated the tournament for 12 years before 1976, winning 10 of 12 championships, making the championship game somewhat anti-climactic. UCLA's coach for that period, John Wooden, retired in 1975, and the UCLA dynasty was over. The championship game became more competitive. Fans started anticipating a close, exciting game. Television ratings soared when Larry Bird dueled Magic Johnson in the 1979 championship game. The NCAA tournament became a national institution in the United States. The phrase "Final Four" was coined in 1975. The Xerox machine was becoming more common in offices, allowing easy duplication of the bracket printed in newspapers just before the tournament. The stage was set for the office bracket pool to emerge.

Why did the bracket pool begin in Staten Island? Why not Las Vegas, the center of the legal sports betting universe in the United States? Las Vegas is a tourist town, and tourists rarely stay for the four weeks between placing bets and seeing the outcome of the tournament. Bracket pools made sense in neighborhoods or offices where the participants live and work together. Betting on the win/lose outcomes of sport competitions is illegal in most of the United States, including Staten Island, but some state penal codes carve out exceptions for "social gambling" when the pool organizer does not profit by charging a fee or taking a cut of the pot. The fairly expansive social gambling exception in the New York State penal code led to the toleration of a public bracket pool, open to all, at a business establishment. Even in states where social gambling is illegal or when federal laws apply, bracket pools are rarely prosecuted (Edelman 2017). Edelman recommends limiting entrants to close friends and paying out all entry fees as prizes. After 29 years of continuous operation, Jody's pool was discontinued in 2007 after Jody Haggerty was investigated by the IRS and fined by the New York State Liquor Authority for promoting gambling in an establishment licensed to serve liquor (Daily News 2007). Press reports indicate that the 2006 pool winner had reported his winnings as income to the IRS, leading to investigations. The pool pot had grown to $1.5 million, and Jody was not reporting the payout of the winnings to the IRS as required by regulations.

The tournament is played after the regular college basketball season, which runs each year from October to March. There are about 350 teams in the NCAA Division I. Division I is composed of the larger colleges, and these generally have the better teams. These teams are organized into conferences consisting of 8 to 16 teams each. During the regular season, each team plays about 30 games, mostly with other teams in their conference. Just after the regular season, each conference plays a conference tournament. The winners of all these tournaments get an automatic invitation (called an "automatic bid") to the NCAA tournament. The rest of the tournament field is fleshed out with teams chosen by an NCAA selection committee. These "at-large bids" are awarded to teams that had a good season with many wins but were eliminated during the conference tournaments.

In the course of the NCAA tournament, all but one team is eliminated by losing a game. The tournament starts with 64 teams (not counting the play-in games) and ends after 63 are eliminated in 63 games. The NCAA tournament is organized into four regional tournaments of 16 teams each. The teams are seeded 1 to 16 in each regional tournament. The teams are seeded by the selection committee based on their season performance and a few other considerations. The lower numbered seed is actually the higher ranked seed, typically referred to as the "higher seed." So, the number 1 seeds tend to be the favorites. Figure 1.1 shows how the brackets for each regional tournament are structured. The victors in the four regions, the Final Four, meet in the final two rounds of the NCAA tournament to decide the national champion.

FIGURE 1.1   The bracket structure for an NCAA tournament region.

The selection committee meets in March on "Selection Sunday©" to determine the at-large bids, to assign teams to regions, and to seed each team in their region. The bracket, with all teams assigned to a specific slot, is revealed around 6:00 p.m. on Selection Sunday. The first game of the first round of the tournaments begins around noon on the following Thursday. During those 90 hours (between the revelation of the bracket and the tip-off of the first game) is when the bracket pool game is played, because you can't fill out a bracket until you know how the tournament is seeded and you are not allowed to enter a bracket in a pool after the games have begun. Your goal is to rack up the most points. You get points for each game when you correctly predict the winner of that game. The specific scoring rules vary from one bracket pool to another, but more points are awarded for games in the later tournament rounds in most pools. Your bracket along with your entry fee must

FIGURE 1.2 President Obama's 2015 Men's bracket (Wall 2015). Licensed under CC BY 3.0.

be delivered to your office pool's manager before the games begin. Figure 1.2 shows President Barack Obama's 2015 bracket.

Now comes the waiting and watching to see how your bracket fares against your opponents' brackets. The 32 games of the first round of the four regional tournaments are played over Thursday and Friday, 16 games per day, with two or three teams defeated and eliminated per hour at some points. The second round of 16 games among the 32 victors in the first round is played on Saturday and Sunday. Then there is a breather till the next Thursday, then four more days of tournament play that winnow the field down to just four teams. Then another breather till the next Saturday night when two games are played and only two teams are left standing. Then, after a rest day, the final game of the tournament, the championship game, is played the following Monday night. The last two teams learn their fate, as do all the hopeful bracket pool players who still have a bracket sheet in the running. In some years, all fates are determined by a ball flying toward a hoop in the last split-second of the last game. The tournament champion is crowned, and all the pooled money is distributed to the bracket pool winners who have the best scores on their bracket sheets.

All gambling is illegal in some states. In 1992, Robert S. Plain of East Greenwich High School in Rhode Island was arrested for possession of gambling paraphernalia. He had in his possession bracket pool entry sheets that he was handing out during homeroom. Mary McNulty, his math teacher of all people, reported him to the police. District Court Judge Robert K. Pirraglia ordered Plain to pay $84.50 in court costs (AP 1992). Fortunately for the future of the bracket pool, this conviction did not create a trend in the United States. Some students at Greenwich protested that betting pools involving teachers were common at the school. The bracket pool was on its way to becoming something of a national institution.

The NCAA created the tournament, but it has no love for the peculiar institution of the bracket pool that the tournament spawned. In the NCAA's own words: "Does the NCAA really oppose the harmless small-dollar bracket office pool for the Men's

Final Four? Yes! Office pools of this nature are illegal in most states. The NCAA is aware of pools involving $100,000 or more in revenue. Worse yet, the NCAA has learned these types of pools are often the entry point for youth to begin gambling. Fans should enjoy following the tournament and filling out a bracket just for the fun of it, not on the amount of money they could possibly win" (NCAA 2010). The NCAA is OK with some bracket contests that give prizes but do not require a wager. The NCAA sponsors its own bracket contest at the website bracketchallenge.ncaa.com.

1.3 A NATURAL EXPERIMENT IN ECONOMICS

About 15 years after the first bracket pool at Jody's, Andrew Metrick somehow convinced his thesis advisor, Eric Maskin (winner of the Nobel Prize for Economics in 2007), to allow him to write a PhD thesis at Harvard on topics that included the Jeopardy!® game show and the NCAA bracket pool. Metrick was an avid chess player in high school and developed an interest in games and decision making. Games that involve decisions and monetary prizes can constitute a natural experiment in economics. The constraints and rules of such a game make it somewhat like a well-designed experiment, one where the results are available without the cost of carrying it out in a laboratory. Metrick used data from 24 bracket pools conducted in 1993 for his analysis of strategic behavior. He also wrote a paper on the bracket pool (Metrick 1996).

1.3.1 Simplify the Problem

In his quest to analyze the strategies of pool players, Andrew Metrick had a problem: there were a lot of strategies. In the parlance of game theory, each possible distinct filled-out bracket represents one pure strategy (a pure strategy is one that is completely defined with no uncertainty). The number of pure strategies equals the number of possible tournament outcomes. A basketball game has two possible outcomes, and each additional game doubles the number of possible outcomes for the set of 63 games that are played. It's like the legend of the wheat and the chessboard. In one telling of the story, Sessa, the inventor of chess, had so pleased his master that the master offers Sessa his heart's desire. Sessa asked that one grain of wheat be placed on the first square of the chessboard, two on the second, four on the third, and so on in a doubling progression until the board was filled. His ruler laughed at such a meager prize, only to find that the 64th square alone must hold over 18.4 quintillion grains of wheat. So it goes, with the possible outcomes when 63 basketball games are played, only half as many, over 9.2 quintillion outcomes. Each pool player can choose from over 9.2 quintillion possible brackets. There are over 9.2 quintillion pure strategies.

Metrick proceeded as if he was following the advice from Polya's *How to Solve It*: "If you cannot solve the proposed problem, then do not let this failure afflict you too much but try to find consolation in some easier success, *try to solve some related problem ...*" (Polya 1973). Metrick simplified the rules of the bracket pools that he was analyzing. There were just two rules:

Rule 1: Pick one winner at random from all players who correctly chose the tournament champion. This amounts to taking all the brackets that picked the champion correctly, throwing them in a hat, and blindly picking one winner.

Rule 2: If nobody qualifies under Rule 1, then pick one winner at random from all the entrants in the pool. That is, throw all the brackets in the hat and pick one winner.

The rules only use the player's pick for champion. The player's down-bracket picks have no bearing on who wins the pool. But these simplified rules retain one important characteristic of actual pools: one player who correctly picks the champion usually wins the pool. Typically, more points are awarded for picking the champion. The most common pool scoring system has the points awarded for each game doubling with each round, awarding 1, 2, 4, 8, 16, and 32 points for tournament rounds 1, 2, 3, 4, 5, and 6, respectively. Under this scoring system and other common pool scoring systems, the winner of the pool is almost always someone who picks the champion (assuming anyone picks the champion). If the champion pick is not decisive in determining the pool winner, then the winner of the pool will be decided before the championship game, making the championship game anticlimactic as far as the pool is concerned. Making it anticlimactic is not necessarily bad, but pool managers tend to set up rules that make this an uncommon occurrence.

Under the simplified rules, there are only 64 pure strategies, corresponding to the 64 teams that a pool player could pick as champion.

1.3.2 A Team's Chances of Winning

Metrick wanted to know if there were any profit opportunities available to players in bracket pools. He had all the 1993 data from 24 bracket pools. So, he had information on the proportion of players that bet each team for champion. He also needed information on the probability that any given team would win the tournament. He got this information from the Las Vegas odds. The bookmakers take bets on the champion and they adjust the odds so that they will have to pay out the same amount of money whether the team wins or loses. Say a bookmaker gets a bet from Larry for $100 that Duke will win and a bet from Moe for $70 that Duke will lose, at even odds. The bookie will charge a fee (or vigorish as it is called, or vig for short) from each bet, let's say 10%. So, the bookie will have 100 + 70 + 17 = $187. But if Duke wins, then he will have to pay out $200 to Larry! The bookie will lose money. Ironically, bookies are not gamblers, at least not in their role as bookies. They are functionaries in a business that will not be viable without a reliable profit. So, the bookie will go to Curly (or anyone else who is interested in betting) to see if he can find someone to bet $30 against Duke. And, he may have to incentivize Curly by giving better than even odds on Curly's bet.

The odds will shift to equalize the amount bet on each side of the proposition

> The betting market futures provide estimates of the win probabilities of the possible champ picks.

So, the bettors cause the odds to shift. In effect, the determination of the odds is crowdsourced. The odds are determined by a market mechanism. On average, sports gambling is a losing proposition because of the vigorish. The bookmakers make a profit and the profit equals the gambler's average losses. But presumably the gamblers make a rational attempt to avoid bad odds and seek good odds, and, in this push and pull, the market ends up producing an estimate of the probability that each team will win the tournament.

1.3.3 Are There Favorable Strategies?

Table 1.1 presents the eight teams from Metrick's study with the highest probability of winning the tournament. The bulk of pool players, 78.6%, chose a 1 seed for champion. Yet, the probability of a 1 seed winning the pool (according to the Vegas odds) was only 37.3%. Among the next four teams, Seton Hall looks to be an outlier, picked much more than the other 2 or 3 seeds listed. This is

TABLE 1.1   Betting Behavior vs. Win Probability in 1993 Bracket Pools

Team (Seed)	Pool Player Behavior: Share of Champion Picks (%)	Win Probability (%)
UNC (1)	22.6	10
Indiana (1)	15.8	9.1
Kentucky (1)	18.7	9.1
Michigan (1)	21.2	9.1
Arizona (2)	1.5	7.8
Duke (3)	5.4	7.8
Seton Hall (2)	10.3	6.9
Cincinnati (2)	0.7	6.1

Source: Data reprinted from Metrick (1996), pages 163 and 166, Copyright 1996, with permission from Elsevier.

because a good proportion of Metrick's data came from New York City where Seton Hall is a hometown team. Almost 16% of players in New York City pools picked Seton Hall for champion, whereas less than 7% picked Seton Hall in the other pools.

Consider picking UNC versus picking Arizona in a pool with 100 players (each playing 1 bracket) under Metrick's simplified rules. If you pick UNC, then you have a 10% chance of ending up in the hat under Rule 1. But around 22 other players will be in that hat, so your chance of being picked at random from the hat is less than 1 in 20, less than 5%. Multiply that by the 10% chance of being in the hat in the first place and you have only a 0.5% chance of winning, less than 1 chance in 200. If the entry fee for the pool is $1, then there will be $100 in the pot. If you win once in every 200 years, then you only average 50 cents per year. Your average return is about 1/2 of your entry fee.

Now consider picking Arizona. You have a 7.8% chance of ending up in the hat. But there will typically only be one or two names in the hat, including yours. Even with two opponent names in the hat, you have a 1/3 chance of winning. Overall, you a have a 7.8/3 = 2.6 or a 2.6% chance of winning the pool. You have a 2.6% chance of winning $100. So, your expected return from the pool will be $2.60 for the dollar you invested in the pool entry fee.

Picking a contrarian champ reduces the competition from opponents that have the same champ.

This is not the more detailed analysis that Metrick performed, but it's a good first cut at estimating the return on investment. One difference is that this example used the average pick share for all pools. Metrick found that the players in larger pools were prone to pick the favorites, like UNC, less than the average player. Hence, those who picked UNC in a pool of size 100 got a return of 80 cents on the dollar, better than the 50 cents or so, but still a losing proposition.

Rather than just using the raw pick share numbers (like 22 for UNC) to directly represent the number of names in the hat, Metrick treated them as point estimates of pick share probabilities. If the raw data showed that 22% picked UNC in a pool, then he estimated that each player had a 22% likelihood of picking UNC. For a pool of size 100, the number of UNC picks could, in principle, vary from 0 to 100, but those extreme values are unlikely. This is like 100 flips of a biased coin with a 22% likelihood of landing on heads. This is called a binomial probability distribution. The outcome will center around 22. Twenty-two will be the average number of UNC picks. The number of opponent names in the hat, if UNC wins, will be between 17 and 27 about 80% of the time. If only 1.5% picked Arizona, then the pick distribution for a pool of size 100 would center around 1.5. The number of opponent names in the hat if Arizona wins would be less than three about 80% of the time. Metrick estimated a $4.39 return for Arizona in a pool of size 100, quite a bit more than my seat-of-the-pants estimate.

1.3.4 Pooled Betting

The sort of pooled betting used in bracket pools is called pari-mutuel betting. *Pari mutuel* is just the French term for pooled betting that found its way into English. Pooled betting has interesting implications for your choice of a champion. The likelihood that you will win the pool depends on how many of the other pool players choose that same champion. If you are the only one to correctly pick the champion, then you win the pool; it's a 100% sure thing under the simplified rules. But if 10 other players choose that same champion, then there are 10 brackets in the hat and you have a 10% chance of winning or a 0.10 probability of winning. However, this has to be balanced against the fact that the popular picks for champion tend to be the most likely winners. This is the bracket pool player's dilemma. The pool player has to balance these two factors: (1) the probability that their champion pick will win the tournament, (2) the probability that the pool player will score enough points in the earlier rounds to beat any opponents who picked the same champion. Or (under Metrick's simplified rules) the probability that their bracket will be picked from the hat under Rule 2.

Horse racing and some lotteries use pari-mutuel betting. Economists had analyzed the strategic behavior of bettors in those other games before Metrick analyzed bracket pools. In horse racing, bettors evidence a long-shot bias; that is, they over-bet the long shots. Lottery players over-bet certain lucky numbers (such as birthdays). So, a smart bettor might be able to gain an edge by betting the favorites in horse racing or avoiding the lucky numbers in the lottery. But in practice, the edge is typically not there because a fraction of the total bets is taken by the house. (Office bracket pool organizers almost never take a cut of the pot because this removes the real or de facto social gambling exclusion and could lead to prosecution.)

Metrick found the opposite of a long-shot bias in bracket pool betting. There is a bias in the direction of the favorites. Why this difference? Metrick speculated that this was due to the fact that the odds offered are explicit in horse racing. The odds are right there on the racetrack tote board, visible to all. In the bracket pool, the odds offered are more obscure. Lottery picks show something similar to a "favorites bias" where birthdays and "lucky" numbers are over-bet. The lottery odds offered for specific numbers are also obscure. It takes some investigation to estimate the odds offered in a bracket pool, some awareness and analysis of past patterns of opponent play. Few pool players evidence any awareness of the odds offered. Most bracket pool players act like someone who goes to the racetrack and plays the ponies without looking at the odds on the tote board.

1.3.5 Some Pool Players Already Knew

Metrick found that there was less over-betting of the 1 seeds in larger pools. The fact that players in larger pools were significantly less prone to pick the favorites indicates something interesting. As Metrick puts it, "a small number of players are changing their behavior to reflect the changing strategic situation." Some pool players had already discovered the gist of Metrick's findings and were playing the pool for a profit. But in spite of the few players that deployed an effective profit-oriented strategy, there were still profits to be made. Metrick estimated that in pools of size 200 betting, Arizona still returned $5.23 on each dollar invested.

1.3.6 What a Competitive Pool Would Look Like

Metrick calculated the overall optimal betting strategy assuming all players had the goal of maximizing their wins/profits. He did this using a "what if" process of progressively shifting player's champion picks to whatever profit opportunities existed until there were no more profit opportunities. This produced an "equilibrium" distribution of champion picks. The equilibrium is where no single player could improve his expected payoff. This is called the Nash equilibrium. John Nash, whose life story was presented in the movie *A Beautiful Mind*, won the Nobel Prize for development of this equilibrium analysis for non-cooperative games.

At equilibrium, the pick percentage of each team is approximately the same as the win probability expressed in a percentage. The deviations from this rule occur in the smaller pools. The smallest pools that Metrick analyzed had 25 players. The deviations were partly due to the fact that there is no profit in betting a team with a win probability of less than 1/25 even if you are the only player who bets them, because your expected payout will be less than 1/25th of the sum of the 25 entry fees in the pot; less than $25*1/25 = 25/25 = 1$ entry fee. So, only nine teams (those with a win probability greater than 1/25) could be profitably bet in a pool with 25 players.

> Don't go too contrarian. Don't pick a champ that has a win probability of less than 1/N of winning the tournament, where N is the total number of brackets in your pool, if you can avoid it.

As the pool gets larger, a larger fraction of teams is profitable to bet. In pools with 200 entries, more than 12 teams were profitable to bet and 16% of the bets went to 4 seeds or lower. (Metrick did not report the details beyond the 3 seeds.) As the number of pool players grows to infinity, the fraction of picks for each team approaches the win probability for that team.

In calculating the Nash equilibrium, Metrick used the pick distributions from the pools. But this detailed information about opponent play would typically not be available when the brackets were being filled out back in 1996. It would have been difficult to arrive at a detailed quantitative estimate of the payoff for betting a specific team. But the fact that 1 seeds would be over-bet could be determined from the pattern of play in past pools. By 2002, the large online bracket contests sponsored by Yahoo and ESPN started posting the pick distributions for all teams in those contests days before the submission deadline for brackets. This data provides a useful estimate of opponent play in office pools except for the hometown effect. If there is a hometown team in the tournament, then it tends to be over-bet relative to the large nationwide bracket contests.

1.3.7 Are Most Pool Players Irrational?

The actual pick distributions were far away from the rational profit-oriented equilibrium. Are pool players just irrational? One viewpoint in economics is that everyone is rational and that some find value in something other than monetary (or pecuniary) profit. Another viewpoint, going under the name "behavioral economics," posits that we are indeed sometimes irrational in our decision making.

Much of standard economic theory is based on the notion that people are a collection of rational agents seeking to maximize profit. Sometimes the notion is just a simplifying assumption that allows progress in an abstract analysis. But it can also function as an invalid assumption in real-world decision making. Metrick's paper on the bracket pool was part of an effort by economist and psychologist to better understand the non-rational component of people's economic behavior, an effort that came into prominence in the latter half of the twentieth century. This effort is important because it can help improve the effectiveness of organizations and decision making and perhaps even help us find better ways to prevent the instability that on occasion plagues the whole economy, as in the severe 2008 recession.

Metrick also calculated a different kind of Nash equilibrium. For this calculation he assumed (for the purpose of the analysis) that all play was rational but that some pool players were getting some kind of non-pecuniary payoff, a payoff in something other than money. A player might simply enjoy correctly picking the correct champ even if they do not win the pool. He considered the case of a player betting for 1 seed Michigan instead of 2 seed Arizona in a 50-player pool with an entry fee of $5. This is one of the hardest decisions to justify because Michigan's win probability exceeded Arizona's only 1.3%, yet Arizona's expected payoff exceeded Michigan's by $11.55. He found that, under this assumption, the player must be getting almost $890 worth of non-pecuniary utility simply from picking the winner. (The $890 figure seems large since the pool pot is only $250. The large figure is due to the fact that the player is forgoing a significant amount of expected pecuniary payoff for a very small 1.3% chance of actually getting the additional non-pecuniary payoff of successfully picking the champion.) Of course, the enjoyment from picking the ultimate winner might not be the only benefit. But the $890 figure gives an idea of the magnitude of the seemingly non-optimal pool betting behavior that needs to be explained.

Figure 1.3 represents a diagram of Metrick's paper. The betting market futures and the competitor's brackets provide sufficient information for two different financial analyses, two different Nash equilibriums. One analysis yields a what-if picture of what rational financial behavior based on a pure dollars-and-cents profit motive would be. The other analysis yields a set of estimates for the value that each competitor is apparently assigning to their

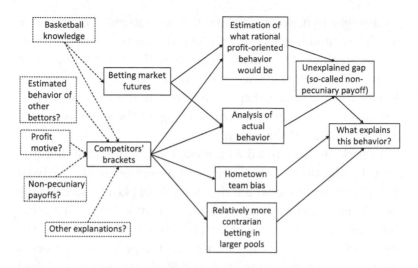

FIGURE 1.3   Key concepts in Metrick's paper.

bracket sheet. There is a gap between the two. This gap is typically given the pro forma name "non-pecuniary payoff." But the term is theory-laden. The term assumes an explanation based on the bettor's value system and avoids the assumption that the bettor's behavior is, at least to some extent, irrational. The gap is really just an unexplained gap. The dotted line boxes indicate some possible influences on the competitor's behavior.

At one point in his paper, Metrick referred to players who push the pools toward the profit-oriented equilibrium as "smart," seeming to imply that their opponents are dumb. That's one theory that could help explain the gap. Three years earlier economist Richard Thaler referred to a large proportion of stock market investors as "blockheads" (Thaler 1993). Tendencies in economic behavior that appear to have no rational basis are referred to as behavioral anomalies. Some behavioral anomalies may exist due to the fact that our brains evolved in an environment where solving problems like filling out the best bracket or solving modern financial puzzles was not important.

Metrick provides a number of explanations of how seemingly suboptimal player behavior might be rational because it made playing the pool more enjoyable. The hometown effect (where Seton Hall gets a higher fraction of picks in New York City, and 6 seed Berkeley was picked by seven players in San Francisco and none elsewhere) might be due to players getting more enjoyment from betting for their favorite team. Also, players might pick objective favorites because they get enjoyment from remaining in contention as long as possible.

> Don't pick the hometown favorite for champ.

Racetracks provide explicit payoff odds. But the payoff odds are more obscure in the bracket pool. Even the raw data that could be used to calculate the payoff odds are unknown since each player's bracket is not revealed to other players until after the deadline for making bets. The odds would have to be inferred from patterns observed from past years' pools. It takes time and effort to do that. The cost of time and effort can be viewed as diminishing the potential profit.

Metrick pointed out that the profits come with the risk of loss. Arizona had only a 7.8% chance of winning. The player who routinely plays a strategy like that will get less than one win per decade on average even in a small pool with 25 players. Much of this financial risk could be diversified away if it were possible to play multiple entries in a pool. Metrick did not consider this, perhaps because it was against the pool rules or otherwise not allowed. Metrick did consider playing different pools, which provides even more effective diversification. Different champions could be bet in different pools. But Metrick pointed out that it was not possible to "poach" office pools all over the nation. Most office pools were closed to outsiders. A few players could not aggregate and capture all the potential profits. It was not feasible to create a bracket pool hedge fund in 1996. This is an important feature that keeps office bracket pool play from being forced to the conventional financial equilibrium by a few profit-oriented rational players.

1.3.8 Was the Simplified Problem Good Enough?

This is all premised on the idea that Metrick's simplified rules are close enough to reality. The simplified rules assume that the winner of the pool always picks the correct champion. Metrick found that all 24 pools he analyzed would have been won by a UNC picker if UNC had won and the rules were the steep points-per-round progression of 1, 2,4, 8, 16, 32 for rounds 1, 2, 3, 4, 5, 6, respectively. Even with the flatter, more gradual, points-per-round progression of 1, 2, 3, 4, 5, 6, 18 of 24 pools would have been won by a UNC picker. Among the 24 pools, eight of the pools uses the steep progression, at least one used the flatter progression, and the median progression was 1, 2, 3, 5, 9, 15. So the assumption that the winner of the pool picks the champs often holds true, particularly for pools that award a relatively large number of points for correctly picking the champion.

1.3.9 Down-Bracket Chalk

To further defend the idea that picking the correct champion is usually key to winning, Metrick suggests a particular down-bracket strategy: he suggests that a player who correctly picks a contrarian champion could defend himself from getting overtaken by a player who scored more points in the earlier round games by simply picking objective favorites in all other games that were not constrained by his championship pick. Picking all favorites would maximize expected points in the earlier rounds. This makes it less likely that any player who picked a losing champion could overtake a player who picked the winning champion by winning enough points in the earlier rounds to compensate for the championship round points.

"Chalk" is a slang term for picking the favorites that originated in the days when bookies would write the horse racing odds in chalk on a blackboard. The odds on the favorites tended to have to be updated more often, so the spot where they were erased and rewritten showed more chalk dust. Hence, a "chalk horse" was a favorite. The first use of the term "chalk bracket" turned up in 2008 when the outcome of the NCAA tournament was the closest it has ever been to a chalk bracket, with all four 1 seeds making it to the Final Four that year.

The goal of Metrick's paper was not to give bracket advice, but the profit-oriented player can derive a likely favorable strategy from Metrick's analysis: Pick a 2 seed that is not a hometown team for champion and go with the favorites everywhere else. Fill out a bracket that is 100% chalk except for advancing one team to the championship that is likely to be under-backed in your pool.

> A down-bracket with no upsets provides you with at least an average chance of coming out ahead of opponents that pick the same champ as you picked.

Metrick proposed the all-chalk down-bracket merely as a simplifying assumption. He assumed it was average versus opposing down-brackets. More recent analysis has shown that an all-chalk down-bracket is typically above average. Hence, in situations where Metrick's analysis recommends a 2 seed for champ, more recent methods may find that a less contrarian 1 seed provides the higher return on investment.

1.4 OBAMA'S BRACKETS

Andrew Metrick served on President Obama's Council of Economic Advisors from 2009 through 2010. But it's clear that Obama did not use advice from Metrick on how to fill out a bracket. Obama's champion picks were too chalky and his down-bracket picks were not chalky enough. When it comes to filling out a bracket, Obama is more of an everyman than a strategic player who is optimizing his chances of winning.

Obama is an avid basketball player and fan. He filled out a bracket in his first March in office, and he assented to ESPN's request to make this a public event. Predictably, Obama got political flack from a number of quarters for filling out a bracket. Duke's coach, Mike Krzyzewski, had this to say about

Obama's 2009 bracket picks: "Somebody said that we're not in President Obama's Final Four, and as much as I respect what he's doing, really, the economy is something that he should focus on, probably more than the brackets" (Fox News 2009). Obama had Duke, a 2 seed, playing to seed expectations, but Obama also picked Duke's archrival, UNC, to win the tournament. Perhaps Krzyzewsky, known to be a shrewd coach, was trying to fire up his team. Obama was criticized for ignoring the women's bracket in 2009. There were complaints about the male-dominated work environment at the White House and the all-male pick-up games on the White House basketball court. Obama filled out a women's bracket every year after 2009 (Wolff 2016). Figure 1.4 shows Obama's 2015 bracket for the women's tournament.

According to the manager of the White House bracket pool, the pool had hundreds of entries (Tracy 2017). Every year, Obama played one of the top two most popular 1 seeds for champion.

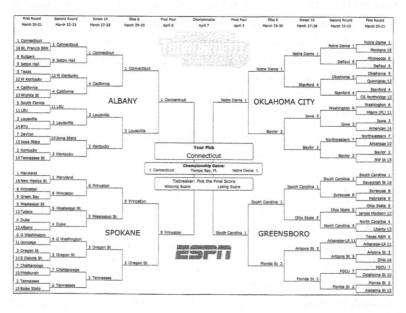

FIGURE 1.4   President Obama's 2015 Women's bracket (Wall 2015). Licensed under CC BY 3.0.

Even if one of his champion picks won (as UNC did in 2009) he would have faced lots of competition. The buzz in the media was that his down-bracket picks were relatively chalky, but not chalky enough by Metrick's standards. He picked 52 seed upsets in the first round and 47 seed upsets in the later rounds over eight years of men's tournament brackets. He had good results on his first-round upset picks, going 28–24. But in the later rounds, where correct picks typically are rewarded more points, his upset picks hurt his brackets; He went 13 for 47 (Mather 2017). Overall, he would have improved his chances of winning if he had just picked the higher seed in all of his men's brackets. In the women's brackets, Obama had a fondness for Princeton in the later years; his niece Leslie Robertson was a starter on the team. He advanced 8 seed Princeton to the Final Four in 2015, but they played to seed expectation and lost to 1 seed Maryland in the second round.

However, the White House "pool" did not require an entrance fee, probably as a nod to federal regulations against gambling in the workplace. It was a contest for bragging rights, not really a betting pool at all. So, all payoffs were non-pecuniary. And Obama surely got some enjoyment from some of his upset picks. One of his riskiest picks was Hawaii to win the first round of the 2016 men's tournament. Obama grew up in Hawaii and he got to go to one of their games with his grandfather when he was age 10, so they were a sentimental favorite of his. And his pick panned out: 13 seed Hawaii beat 4 seed California 77–66. Also, even if Obama had optimized his brackets, it's still more likely than not that he would have never won any of the eight pools he played as president. In a pool of size 200, a middling player will win once every 200 years on average. An optimized bracket that is five times better than average will still only average one win every 40 years.

1.5 METRICK'S IMPACT

There is every indication that word did not get around quickly about Metrick's analysis. The pundits in the media kept advising pool players to pick lots of upsets in the early rounds and to pick a champion with no thought about their opponent's picks.

In 2005, the *New York Times* published a piece that referenced Metrick's findings that the No. 1 seeds are over-bet (Leonhardt 2005). Since 2005, the notion of betting a contrarian champion while picking few or no upsets in the early rounds has gotten somewhat more attention.

In 2015, Metrick gave a rare interview on the bracket pool to the *Wall Street Journal* saying, "It often pays to be contrarian, you want to take good bets that others don't want: stocks that are out of style for behavioral reasons and basketball teams that are good but may not have the most fans" (Cohen 2015).

The *Journal* even managed to get a quote on the bracket pool from John Nash. In an email, Nash wrote "I guess I am myself pretty far from being the sort of person who would 'fill out an NCAA tournament bracket'."

CHAPTER 2

Predicting the Tournament Outcome

METRICK FOCUSED HIS ANALYSIS on picking the winner in the last round at the top of the bracket. The other 62 down-bracket games were just used to support this champ pick. He assumed that simply picking all favorites in those 62 games was the best way to defend his bracket from being outscored by a wide margin in the earlier rounds. This method will get you a good bracket (if not always the best bracket) in most pools because the scoring in most pools is top-heavy: typically, a relatively large number of the points can be scored by just picking the right champion. But an analysis of the whole bracket is necessary to better quantify how well the down-bracket supports the champ pick and to determine when one or more down-bracket contrarian picks are an improvement over Metrick's all-favorites down-bracket strategy.

Metrick used two basic models as inputs in his analysis: (1) a *tournament outcome model* that assigned advancement probabilities to teams in the tournament (2) and *opponent model* that assigned advancement probabilities to the champ picks of opposing players in the office pools.

Metrick used a tournament outcome model based on betting market futures for the tournament champion and just assumed that the higher seed was the most likely winner in earlier rounds. The goal of this chapter is to extend the tournament outcome model to all rounds of the pool. Chapter 5 will address the opponent model.

2.1 AN EXAMPLE OF A TOURNAMENT OUTCOME MODEL

Figure 2.1 shows the bracket for a four-team, two-round tournament. Table 2.1 shows an example of a tournament outcome probability model for this tournament. The rows represent the win probabilities for each team. The columns represent each of the team's opponents. For instance, row A represents the win probabilities of team A against the other three teams. Scanning the probabilities, you can see that the teams are ordered by relative strength, A>B>C>D in terms of win probability. Typically, the first round of a tournament matches the best versus the worst,

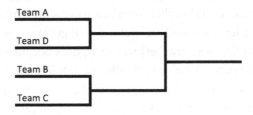

FIGURE 2.1   A four-team tournament bracket.

TABLE 2.1   A Tournament Outcome Model

	Team A	Team B	Team C	Team D
Team A		.57	.70	.78
Team B	.43		.64	.73
Team C	.30	.36		.60
Team D	.22	.27	.40	

From Niemi et al. (2008), copyright © American Statistical Association, www.amstat.org, reprinted by permission of Taylor & Francis Ltd, http://www.tandfonline.com on behalf of American Statistical Association.

the second-best versus the second worst, and so on. This gives the best teams a better chance to survive until the later rounds and rewards the best teams for playing well during the season before the tournament by giving them easier early round games. The best teams already have a leg up even before the first round starts. The NCAA has not always accomplished the goal of ordering or seeding the tournament from best to worst for various reasons that will be discussed later, but we will assume the ideal seeding in this example. So, A will play D and B will play C in the first round.

There are 16 cells in the matrix of probabilities for a four-team tournament. A team's row lists its probability of winning against each opponent. A team's column lists its losing probabilities. The four diagonal cells are empty because they represent a team playing itself. That leaves 12 probabilities, but six of those cells can be derived from the other six. When A plays D, one of them must win the game. There are no ties in basketball; overtime periods are played until one team wins. It's a 100% sure thing that either A or D will win the game. Hence, the probability of one of the two teams winning the game is 100% or 1.0. The set of all possible outcomes is called the *sample space* and the probabilities of all possible outcomes must sum to the value 1.0. So, if the probability of A beating D is 0.57, then we can calculate the probability of D beating A as 1.0 − 0.57 = 0.43. In general, the probability of one team winning a game is just one minus the probability of the other team winning. This rule holds for all the first-round matchups. So, we really only need to know six probabilities to create the outcome model for a four-team tournament. For the full 64-team tournament, we would have a table with 64 rows and 64 columns, for a total of 64 × 64 = 4096. But 64 of those are blank cells on the diagonal and half the remaining cells can be calculated from the other half. So, we need to estimate 2016 game outcome probabilities to create a tournament outcome model for a 64-team single elimination tournament.

When Tim Trowbridge and his buddies got together in a Kent, Ohio bar to plan their first bracket pool, their goal was to create a definitive test of basketball knowledge. In all the analyses that follow, this set of 2016 game outcome probabilities boils down all the basketball knowledge needed to estimate your best picks in your bracket pool.

2.2 WHY USE A PROBABILITY MODEL?

Most people don't use an explicit tournament outcome model to make picks in a bracket. But they do make some implicit use of approximate probabilities. They will pick all the 1 seeds to win in the first round because they know that 1 seeds are very likely to win. People are indecisive about whether to pick 8 seeds or 9 seeds because the win probabilities are close to 0.5. They think it is necessary to pick some upsets in order to win a bracket pool. Picking an upset means picking the team with a win probability of less than 0.5.

We will make more explicit and quantitative use of probabilities. Metrick used the estimated win probability of each team along with an estimate of the number of opponents that picked the same team to determine the return on investment for each team. Some pools offer upset incentives for picking upsets. For example, a pool might offer 13 points for picking a 13 seed and only 4 points for picking a 4 seed. If the win probability for a 4 seed is 0.8 and you pick the 4 seed, then you will get 4 points 80% of the time and zero points otherwise. So, the average or expected number of points you will get is $4 \times 0.8 = 3.2$. If you pick the 13 seed, the expected points will be $13 \times 0.2 = 2.6$. The tournament outcome model is useful because we can use it along with other estimates (like the estimated number of opponents that will pick a team or the points awarded for correctly picking a team) to estimate values like return on investment or the expected score that a bracket will get. We can get quantitative estimates that are important to our goal of improving our brackets.

2.3 ESTIMATING GAME OUTCOME PROBABILITIES

The six required game outcome probabilities in the example tournament outcome model in Table 2.1 are just made up numbers. We want good estimates for the probabilities required to create a real tournament outcome model because this model is going to be part of the basis for our decisions as to how to fill out our bracket. We want the best estimates we can come up with. How are we going to arrive at these estimates for the outcome of each possible game in a tournament?

Biostatistician Brad Carlin (Carlin 1996) developed a method that can be used for estimating the game outcome probabilities for all possible matchups in the full tournament bracket. His method was the first to use team-specific information to estimate the game outcome probabilities for the NCAA tournament. An earlier method was based solely on the tournament seedings. But there are only 16 distinct seeds. Sixteen sets of 4 seeds are assigned the same seed. So, all the 1 seeds for all tournament years will have the same strength if only the seeds are used. Carlin assumed that a more fine-grained distinction between team strength would be better. His method uses team strength information from betting market data for the first round of the tournament and Jeff Sagarin's computer ratings (Sagarin 2017) for the later rounds. The general method of converting a point spread to a win probability was originally developed by Hal Stern and first applied to professional football (Stern 1991). It's sometimes called Stern's Law.

2.4 CONVERTING A POINT SPREAD TO A PROBABILITY

Betting on game outcomes is popular, but relatively few bets are placed directly on the odds of victory. Most betting is on the margin of victory called the "point spread." If the point spread is 6.5, then the bettors are betting on the proposition that "the favorite will win by 6.5 or more points." The bettors may bet either side of the proposition. The bookmaker offers the same payoff per dollar bet on either side of that proposition. Bookmakers estimate the point spread of the game and adjust it as necessary, as the bets come in, presumably seeking to have the same amount of money bet on both sides of the proposition. Having even money on both sides of the proposition ensures that they get a reliable profit. The bets by the losers have to cover the payoffs to the winners; otherwise the bookmaker has to use his fees or more to make the payoffs. If the bettors (as a group) think that the spread is below the expected margin of victory for the favorite, then they will bet more for the favorite than for the underdog and the bookmaker will have to increase the spread. If the bettors think the spread is above the margin of victory, then they will bet more for the underdog and the spread will decrease. If the betting is even on both sides of the spread, then there is a consensus among the bettors that the point spread is an accurate prediction of the favorite's margin of victory. Carlin used the point spread as a crowdsourced estimate of the margin of victory. Research cited in Carlin (1996) has shown that betting market point spreads are the best available estimator of the margin of victory in football games. Wolfers (2006) shows that Stern's Law performs well on a data set consisting of over 70,000 NCAA games in spite of some identified bias in the betting market point spreads relative to the point margins of the actual games. (Wolfers identified point spread bias in college basketball and attributed the bias to point shaving, but other analyses [see Borghesi 2015] suggest that non-corrupt game management practices by coaches are a factor.)

Betting market point spreads are the best predictors of the first-round winners.

Carlin needed a way to convert all the point spreads to probabilities. This is easy if the point spread is zero, the probability of victory for either team is 0.5. But what about other point spreads? Carlin solved this problem by using one of the most important tools in a statistician's toolkit, the *central limit theorem*. The central limit theorem states that the outcomes of most random processes deviate from their mean value in accordance to the normal distribution, the bell curve.

Abraham de Moivre discovered a special case of the central limit theorem and published it in 1738 in his book *The Doctrine of Chances or A Method of Calculating the Probability of Events in Play.* He proved that outcome of a series of coin flips, the number of heads, was distributed according to the bell curve. In one example, he calculated that for 3600 coin flips the probability that the number of heads being within 30 of the mean is .682688 or 68.2688%. This value is what we now call the *standard deviation* of the normal distribution.

De Moivre's special case of the central limit theorem can be demonstrated with a Galton Board (Figure 2.2). Marbles are channeled in at the top of the board so that they hit a series of pins. The pins are precisely located so that each marble has a head-on collision with each pin in its path and a 0.5 probability of falling either to the right or the left of each pin, just as a flipped coin has a 0.5 chance of falling heads or tails. The marbles fall out of the bottom into a series of bins and preserve a record of the sum of their series of small deviations to the right or left. The result is an approximate bell curve made out of marbles.

Carlin used de Moivre's trick. Carlin's method for converting a point spread of 11 points to a probability is illustrated in Figure 2.3. The mean of the bell curve falls at 11, this is the crowdsourced estimate for the average margin of victory. The standard deviation of any college basketball game's margin of victory is estimated to be 11 points (Breiter and Carlin 1997). (This 11-point estimate is based on observations like those presented in Figure 2.5 and discussed later in this chapter.) So, the probability of the margin of victory being within 11 points of the point spread is 0.682688, about a 68% chance. There is a 32% chance that it will not be that close to the mean; a 32% chance that the margin of victory will be in one of the tails of the normal distribution. Since the bell curve is symmetric, the two tails are the same size. The two tails are mirror images of each other. So, there is a 32/2 = 16% chance that the margin of victory will be in the lower tail. If the favorite's margin of "victory" is in the lower tail, then it's more than 11 points lower

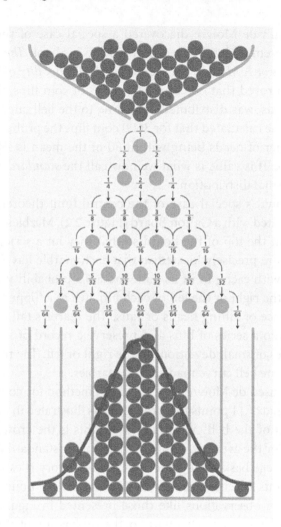

FIGURE 2.2 The Galton box illustrates the central limit theory and the normal distribution. (© Peter Hermes Furian/Dreamstime.com.)

than the spread. That's a negative number. That's a loss. Therefore, the probability that the favorite will lose the game is 16% and the probability that the favorite will win is 100% − 16% = 84%. In this example, we used a point spread that equaled one standard deviation. But, the method can be used to convert any college

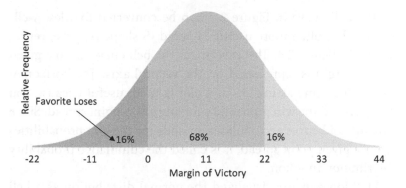

FIGURE 2.3   Converting a point spread to a game outcome probability using the central limit theorem.

basketball point spread to a win probability, on the assumption that the margin of victory is normally distributed with the mean equal to the point spread and a standard deviation of 11. The calculation amounts to moving the bell curve depicted in Figure 2.3 to the right or left till the mean of the bell curve coincides with the spread and then determining the area of the bell curve to the left of zero. You can use the NORM.DIST function in an Excel spreadsheet to do this calculation. For instance, for a spread of 6.5

$$\text{NORM.DIST}(0, 6.5, 11, \text{TRUE}) = 0.277$$

If the point spread is 6.5, then the probability that the underdog will win is 0.277.

The vertical axis in Figure 2.3 is labeled "Relative Frequency." But we are trying to predict the probability of a single game that will have only one margin of victory. In actuality, there will be no relative frequency of outcomes. We are using a *frequentist* interpretation of the probability here. We are imagining what would happen if this game with a point spread of 11 was played over and over again. The frequency of losses would be about 16% of the games played. This is perhaps the simplest interpretation; the one that is the easiest to understand. But, granted, it's a bit awkward, and some philosophers of probability might dispute it.

The bell curve in Figure 2.3 can be converted to a less well-known, but often more useful, sigmoid (S-shaped) curve, represented in Figure 2.4. The area under the bell curve up to a given point margin is represented on the vertical axis. The horizontal axis is the same as in 2.3. This plot is more useful because you can just read the win probability for any given point spread. Since this sigmoid curve accumulates, or adds up, the win probabilities up to a given point spread, it is called the cumulative probability distribution function.

De Moivre never visualized the normal distribution as a bell curve. He worked it out in algebra. There is no evidence that he ever found a practical mathematical application for it outside of gambling. But he did realize the significance of his discovery that the number of heads and the number of tails resulting from a series of coin flips tended toward equality in a measurable manner. The general idea that the number of heads and tails in a series of coin flips tended toward equality was known before de Moivre.

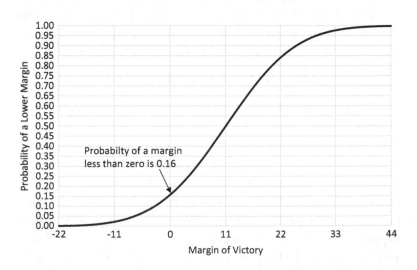

FIGURE 2.4 The cumulative probability distribution function for Figure 2.3. This sigmoid curve shows the area under the bell curve up to any given point.

This is called the "Law of Large Numbers." De Moivre discovered that the expected rate of closure to equality could be calculated.

Of course, it's a bit of a leap to apply a discovery about coin flips to the basketball point spread. But this move has proven to be effective in a wide variety of applications. In the early 1800s, Pierre-Simon Laplace proposed that errors in observations have a normal distribution because they are composed of numerous small independent deviations around the accurate value (Salsburg 2017). The margin of victory in a basketball game results from a series of scoring attempts that either succeed or fail. The central limit theorem has been proven correct for a range of random processes and it's considered to be approximately correct for a wider range of applications. Statistician David Salsburg prefers to call it the central limit conjecture since it is often used for applications that are a bit out of the range of proof (Salsburg 2017). Figure 2.5 shows the deviations from the betting market point spread for actual margins of victory. They are close to the normal distribution with a standard deviation of 11 points, which is superimposed over the frequency chart of game outcomes.

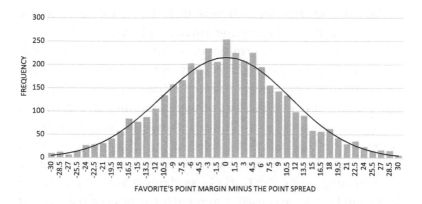

FIGURE 2.5 Deviations of margins of victory from the point spread for the 2017 season, including a superimposed normal distribution with a mean of zero and a standard deviation of 11. (Data from ThePredictionTracker.com.)

2.5 USING RATING-BASED SPREADS

Carlin used point spreads that were derived from the Sagarin ratings after the first round, since there are no betting market spreads available prior to the tournament for games after the first round. The rating-based spread for a game is calculated by subtracting the rating of one team in the game from the opposing team. Then the ratings-based spread can be converted to a probability using the central limit theorem and the standard deviation of 11. But first, Carlin investigated the rating spreads to see if they were good substitutes for the betting market spreads. He compared the first-round betting market spreads with rating-based spreads.

He plotted the rating-based spreads against the betting market spreads and used a method called "least squares" to analyze the relationship between them. The method is called "least squares" because it involves fitting a line (or curve) to the data points such that the sum of the squares of the residuals are minimized. The residuals are the vertical distances between the data points and the fitted line. As an example, a line is fitted to five data points and one of the five residuals is identified in Figure 2.6.

The least squares method was first used to determine the planetary orbits, including the smallest observable planetoids. Without an accurate determination of the orbital curve, it was easy to lose track of these faint objects and have difficulty relocating them later. Astronomers knew that the orbits were elliptical, with the sun at one of the foci of the ellipse. But measurements of the locations of these small, faint planetoids lacked precision. Adrien-Marie Legrande first published the least squares method in 1805 and other mathematicians quickly found it to be effective for estimating orbits. But Carl Gauss claimed to have discovered the method earlier and used it to predict the orbit of the asteroid Ceres. Ceres was discovered as a faint object close to the sun, and it moved behind the sun's glare after only 40 days of imprecise observations. This set up a kind of competition among mathematicians to predict where Ceres would next be seen. Gauss's

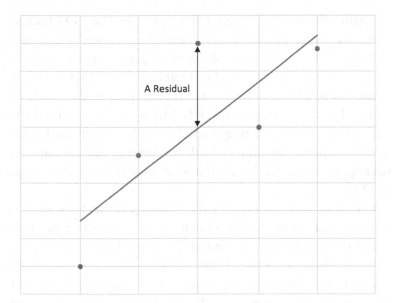

FIGURE 2.6   A residual is the vertical distance between a data point and a line fitted to the data points.

prediction was the most accurate by a wide margin. Gauss claimed to have communicated the least squares method to others, but no one had a clear recollection of that. Gauss also determined that the least squares method was the best of all possible curve fitting methods if the errors in observations have a normal distribution with a mean at the correct value, thereby giving the least squares method a theoretical foundation (Stigler 1981).

Both Laplace and Gauss failed to credit de Moivre for his priority in discovering the importance of the normal distribution in statistics. To this day, the normal distribution is often called the Gaussian distribution and rarely called de Moivre's distribution. De Moivre was ahead of his time. He was an exiled Huguenot working in London. Isaac Newton was among his admirers. But he was unable to get a university post, perhaps due to religious discrimination. He had to work in the gig economy in London, tutoring students and pricing annuities for investors (Bellhouse 2011). He was arguably the first professional statistician.

Carlin plotted the 32 Sagarin spreads for the 1994 tournament first-round games against the betting market spreads and found that the Sagarin spreads were in good linear agreement. He did not need to fit an orbital curve or any other curve, he could just fit a straight line. But the Sagarin spreads underestimated the betting market spreads. The least squares fitted line implied that the Sagarin spreads had to be multiplied by 1.16 to be in optimal agreement with the betting line spreads. (In later applications, Carlin used a more conservative adjustment of 1.05.)

One drawback to using Sagarin's ratings is that they are proprietary, and his method is a trade secret. The ratings are published in *USA Today*. Starting with the 2001–2002 season, the published Sagarin ratings table revealed some clues about the nature of the ratings. There were three ratings: the overall rating, the Elo chess rating, and the Predictor rating. The overall rating was a synthesis of the Elo chess and Predictor ratings. The commentary accompanying the 2001–2002 ratings table says: "In ELO CHESS, only winning and losing matters; the score margin is of no consequence, which makes it very 'politically correct'. However, it is less accurate in its predictions for upcoming games than is the PURE POINTS, in which the score margin is the only thing that matters. PURE POINTS is also known as PREDICTOR, BALLANTINE, RHEINGOLD, WHITE OWL and is the best single PREDICTOR of future games" (Sagarin 2002). (The capitalization is from the original.) The aliases Ballantine, Rheingold, and White Owl arise from the notion that Predictor is the rating system that "beer-quaffing, cigar-smoking sharpies" would be using (Wolff 2003). Sagarin's overall ratings were a synthesis of Elo chess and Predictor until the 2012–2013 season.

Sagarin described the Predictor as the "single best predictor of future games." This seems to imply that the Predictor is a better predictor than the overall Sagarin rating. But why would

Sagarin publish an inferior overall rating? Perhaps it is because Sagarin found that the fans tended to disagree with his ratings (Jonietz 2002). Humans tend to put too much stock in the win–loss record, but the Predictor seeks to make optimal use of the score margin. A team with a perfect record of wins but with an outsized number of close games might rank below a team with a few narrow losses and more blowout victories according to the Predictor ratings. If a rating system is not popular, then it might reduce interest and make the rating system a less valuable commodity. Perhaps Sagarin found it beneficial to balance accuracy against popularity.

But maybe Sagarin meant to say that the Predictor was the best rating among those based on a single parameter like score margin or win percentage. The synthesis of the Predictor and Elo chess could have comparable predictive power, depending on the details of how they were combined.

The overall Sagarin rating may be unbiased, but you probably use biased, crowd-pleasing ratings almost every day. The rain predictions from commercial sources are biased. On average, the Weather Channel predicts a 20% chance of rain when the best prediction is 5%. They do this because if they fail to predict rain and it rains then they get a lot of complaints. If they predict rain and it does not rain, then there are few or no complaints. If you want an unbiased prediction, use weather.gov (Silver 2012).

Shortly after the Predictor ratings became available, Carlin switched to using them for the creation of tournament outcome models. Let's perform a least squares fit of the Predictor spreads against the betting line spreads (see Figure 2.7). The fit indicates that these more recent Predictor spreads must be multiplied by 0.9148 rather than 1.16 to be calibrated to the betting market spreads.

Computer rankings that utilize margin of victory are better predictors than the tournament seeding.

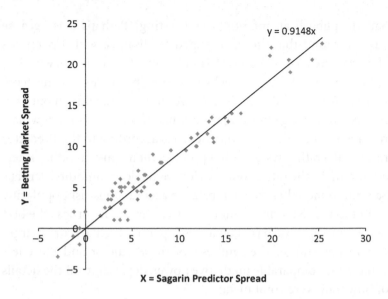

FIGURE 2.7 Predictor spreads vs. betting market spreads for 2018. The line was fitted using least squares. (Data from ThePredictionTracker. com.)

2.6 IMPROVING THE TOURNAMENT OUTCOME MODEL

The Sagarin Predictor ratings do have limitations when used to predict the tournament. The ratings do not take into account injuries, illnesses or suspensions of key players. Ratings that are based only on game score margins are premised on the idea that a team has the same fixed capability throughout the season. But teams missing key players will lose capability. If a key player returns later, then the team's capability will improve. So, ignoring player absences can bias a rating in either direction.

In the year 2000, Kenyon Martin, the best player on the best team (Cincinnati) in college basketball, broke his leg just before the tournament. After losing Martin, Cincinnati's Sagarin rating was about 5 points higher than it should have been, inferring this number from the NCAA tournament first-round betting market spread for Cincinnati's game.

In 2013, ESPN introduced the BPI rating that takes such player absences into account. The original BPI was not designed to be purely predictive. "By reflecting a résumé, BPI was not explicitly built to make predictions" (Oliver 2012). In 2016, the BPI was revised to be a predictive rating (ESPN 2016).

Nate Silver's FiveThirtyEight.com website has started providing a tournament outcome model each year for the men and women's NCAA Division I basketball tournament, and this model takes player absences into account. His overall approach is to aggregate ratings and rankings from a number of sources. Not all these input ratings and rankings are highly predictive, but he presumably weights them so that the aggregated output is unbiased and predictive. However, the FiveThirtyEight.com tournament outcome information is presented as a 6 × 64 table of advancement probabilities for each team for each tournament round, and there is no exact method for reverse engineering this table to produce the 64 × 64 outcome model table of head-to-head probabilities like the one presented in Table 2.1. Still, the outcome model could probably be well approximated.

2.7 PRECISION VERSUS ACCURACY

When I mentioned that the measurements of locations of planetoids in the early 1800s lacked precision, I was using a technical meaning for the word *precision*. You might think that I meant that these measurements where inaccurate. I did not mean that. In the thesaurus, *precision* and *accuracy* are synonyms. In the dictionary, one is used to define the other; one might as well be a one-word definition of the other. But in statistics, the two words have different meanings (Figure 2.8).

Imagine a basketball player is practicing free throws. If the shots are all falling close to the same location, but that location is not the basket, then the shots have high precision and low accuracy. If

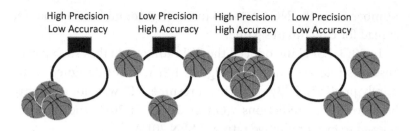

High Precision | Low Precision | High Precision | Low Precision
Low Accuracy | High Accuracy | High Accuracy | Low Accuracy

FIGURE 2.8   Precision vs. accuracy. (Ball image © Scorpion26/Dreamstime.)

the shots are all over the place, but their average location is near the center of the hoop, then the shots have low precision and high accuracy.

If those measurements of the locations of planetoids were inaccurate then the least squares method would not have worked as well. The least squares method is typically effective in the face of imprecision; a curve fit to imprecise measurements will be more precise than each individual measurement. But the least squares method is not, in and of itself, a cure for systematic inaccuracy in the measurements. It can help one identify individual inaccurate measurements that are gross outliers relative to the other measurements.

The Weather Channel predictions are biased toward rain. This is an example of lack of accuracy. In statistical terminology, bias is the opposite of accuracy.

We were able to use the least squares method to calibrate the Predictor spreads so that they could be used as accurate substitutes for the unavailable betting market spreads. But the mathematicians who were estimating the orbit of Ceres had no way to make their orbit more accurate by calibrating the original measurement device. It was too late for that. Fortunately for Gauss, the astronomer had made accurate measurements.

Of course, there is the backboard. If a basketball player is an imprecise free throw shooter, then biasing his shots long (so that some shots might bank off the backboard into the basket) might lead to a higher free-throw percentage. This might be another example of where a biased estimate can be useful. Or a shooter might even aim at the backboard instead of aiming at the center of the hoop. I will let any readers interested in this topic do their own research on the utility of bank shots. It's a complicated subject all its own.

2.8 WILL A 16 SEED EVER BEAT A 1 SEED?

We now have the answer to this question, due to the stunning 16 seed University of Maryland, Baltimore County (UMBC) victory over 1 seed Virginia in 2018. The answer to this question was more of a mystery when Hal Stern and Barbara Mock first tackled it (Stern and Mock 1998). They used three different estimation techniques. The first estimation method they used was the one described here, based on the evidence that the margin of victory closely approximates a normal distribution with a standard deviation of 11. They used the method on the historical average of point spreads of games between 16 seeds and 1 seeds, 26.4 points. That gave an estimated probability of 0.008 for a win by the 16 seed. For the second estimate, they substituted the historical average of all the actual margins of victory, 22.9 points, in place of the average betting market point spread. That yielded a probability of .019. For the third estimate, they used a quite different method called logistic regression on historical win probability versus the seed difference and that yielded a probability of .012. All three estimates were in fairly close agreement. Based on these results, they decided that the probability of a 16 seed beating a 1 seed was about 0.01 or 1%.

They considered a fourth estimation technique but rejected it. The fourth was based on the historical record. That estimate is 0. This implies that a 16 seed will never beat a 1 seed. They judged that as an unsatisfactory estimate because similar things had happened. For instance, 15 seeds had beaten 2 seeds three times and 1 seed Georgetown squeaked by 16 seed Princeton in 1989 by the narrowest of margins, 1 point, by virtue of a missed shot by Princeton at the buzzer.

After the UMBC victory, the historical record gives us a new estimate of the probability as of 2018: 1/136. Stern and Mock's estimate of 1/100 is now closer than before to the observed frequency.

2.9 JUDGING MODELS BASED ON THE TOURNAMENT OUTCOME

A tournament outcome model can be judged on some of the characteristics mentioned in this chapter. Is the model consistent with the betting markets? Does the model account for injuries? But, tournament outcome models are perhaps best evaluated based on the bottom line: How well did the model predict the actual outcome of the tournament?

There are two yearly online contests at Kaggle.com where you can compete on creating the best tournament outcome models for the men's and women's tournament. You don't fill out a tournament bracket. Instead, you fill out a tournament outcome model. These competitions are great for exploring the concept of a tournament outcome model. Before the competition, there is a preliminary phase devoted to developing and testing predictive tournament outcome models.

The Kaggle contest submissions are scored based on the outcome of the tournament like any bracket pool. But, unlike bracket pools, each competing tournament outcome model is not scored based merely on whether the model predicts a win or not. It is scored based on the probability that the model assigns to the win. An outcome model is scored based on the 63 probabilities that the model assigns to the 63 game outcomes that constitute the overall tournament outcome. The 63 probabilities are multiplied together to determine the probability that a competing tournament outcome model assigns to the actual outcome of the tournament.

The game probabilities are multiplied together because that is the way to calculate the joint probability of independent events. If you flip a fair coin, then the probability that it will land heads is 1/2. If you flip two such coins, then the results are independent. The joint probability of both coins falling heads is $1/2 \times 1/2 = 1/4$.

Consider the tournament outcome model in Table 2.1. Suppose the four-team tournament was played and the outcome was that Team A beat Team D and Team C beat Team B in the first round and then Team A beat Team C in the second round. According to Table 2.1, the probability of this outcome is $0.78 \times 0.36 \times 0.7 = 0.19656$. This would be the score of Table 2.1 under the Kaggle contest rules as they would apply to a four-team tournament.

In the real Kaggle contest, the winning score is typically an exceedingly small probability, reflecting the fact that is it exceedingly hard to perfectly predict the outcome of the 64-team tournament. The highest score for the 2017 contest was about 0.000000000001. Scores that tiny are a bit cumbersome to put on a leaderboard, so the natural logarithm of the score is used, and the logarithm is divided by the number of games in the tournament (63) and negated. The actual score on the 2017 Kaggle leaderboard was 0.438575. This is actually the *lowest* score on the leaderboard. This contest is like golf, the lowest score wins. Negating the logarithm of the probability reverses the order of the scores, but it has the advantage of making all the scores positive. This math converts the probability into something called the logarithmic loss metric or log loss. The bigger the log loss, the more inaccurate is the tournament outcome model. Log loss is a useful metric in information theory. But in the context of this Kaggle contest, its only practical function is to provide a way to transform the very small probability scores into manageable values for display purposes. Also, dividing by the number of games normalizes the score so that it represents the average game score. That makes it easier to compare scores for different size tournaments and scores when multiple tournament outcomes are scored as one outcome. Carlin (1996) was the first to use log loss to compare the accuracy of tournament outcome models. He used the log loss metric to provide evidence that the refinements of his tournament outcome model (using the Sagarin ratings instead of the seed values, using the betting market spreads for the first round, and calibrating the Sagarin ratings) increased the accuracy of the model.

2.10 JUDGING A MODEL-GENERATING METHOD

A tournament outcome model is specific to one year's tournament. But the outcome of one NCAA tournament represents a pretty small dataset for evaluating a predictive model. The second stage of the Kaggle contest is a competition for the best performing model for the current year. The first stage of the contest is designed to evaluate model-generating methods. In 2017, the first stage involves submitting tournament outcome models for the past four tournaments. The models have to be based on historical data available prior to the current year's tournament. The first stage competition score is the average score of the four submitted models. A method's ability to generate accurate tournament outcome models can be evaluated based on its performance in prior years.

But Gregory Matthews and Michael Lopez, the team that won the 2014 second stage Kaggle contest, argued that a model should not be tested only on the outcomes of the tournament games. They used the regular season games as well. Hence, their tournament outcome model was built on a more general college basketball game outcome model (Matthews and Lopez 2014).

2.11 USING A TOURNAMENT OUTCOME MODEL IN BRACKET POOL STRATEGY

We could use a tournament outcome model directly to determine our bracket sheet. We could just bet the most likely winner in each game. But that is typically not the best thing to do. Metrick found that picking a contrarian champ was better in most cases. But Metrick did use a tournament outcome model to estimate and optimize return on investment. He used a tournament outcome model along with contrarian considerations to determine which champ pick was best.

But surely submitting the best possible tournament outcome model to stage two of the Kaggle contest is your best shot. No, not if your goal is to win a prize. The rules do not limit you to submitting the best predictive model as determined by stage one. You don't even have to compete in stage one. The organizers of the contest allow you to submit two entries to the stage-two competition. It's a violation of the rules for competing teams to collude and the organizers do some analysis to detect collusion. But it's perfectly OK for you to collude with yourself on your two submissions.

Before describing how you can collude with yourself, we need to consider some characteristics of the scoring system. The 63 game probabilities are multiplied together to get the overall tournament outcome probability. But the logarithms of the game probabilities are added to get the logarithm of the overall tournament outcome probability. If a tournament outcome model assigns a probability of 0.5 to an 8 seed beating a 9 seed in the first round, then this will add $-\log(0.5)/63$, about 0.01, to your score regardless of whether the 8 seed or 9 seed wins. If the model assigns 1.0 to the 8 seed beating the 9 seed, then this will add $-\log(1.0)/63 = \text{zero}$ to your score when the 8 seed wins and $-\log(0.0)/63 = \text{infinity}$ to your score when the 8 seed loses. Since lower scores are better, assigning 1.0 to a game outcome is risky. But if you could manage to assign 1.0 to all the 63 correct outcomes, then you would have a perfect score of zero in the Kaggle contest. However, you would get an infinite score, the worst possible score, if you were wrong about the outcome of only one of the 63 games.

Since the contest rules allow two submissions, you can risk an infinite score on one of your entries in order to get an improved score on the other. Pick a game that is close to a toss-up, like one of the first-round 8–9 seed matchups. In the best possible predictive tournament outcome model, the 8 seed would be assigned a win probability close to 0.5. But you could assign that 8 seed a probability of 1.0 in one of your Kaggle submissions and assign it a probability of 0.0 in the other submission, while leaving all other probabilities at their best possible estimates. You are certain to get exactly the right answer for this game in one of your submissions. Hence, one of your submissions will outperform the best possible tournament outcome model. Assigning 0.5 to the win probability adds about 0.01 to your score regardless of whether the 8 seed or 9 seed wins that game. But one of the two modified models will add zero to your score for this game because one assigns a probability of 1.0 to the game outcome. This strategy is called arbitrage. A score decrease of 0.01 can be significant in the Kaggle competition. In 2017, that score decrease would have moved you from 15th to 5th, and 5th place wins a prize. But the prize was just a hat, t-shirt, and laptop sticker in 2017. In 2016, the 5th place prize was $2000, and this arbitrage strategy would have moved a 12th place entry into 5th place. There is evidence that some competitors were using this arbitrage strategy as early as 2015.

There are other Kaggle strategies that involve tweaking a predictive outcome model. A predictive model won the first year (2014) of the contest. But the 2015 winning model assigned a 1.0 probability to the proposition that 14 seed Georgia State would beat 3 seed Baylor (Sonas 2015; Bradshaw 2015). The 2015 winner was obviously not using a predictive model for one of his entries. There was about a 20% chance of Georgia State beating Baylor. The submitter of the model took an 80% risk of getting an infinite score on one of his entries for a 20% chance of getting an advantage of about $-\log(0.2)/63 = 0.025$ over predictive models. The submitter won the $10,000 first prize. The submitter was Zach Bradshaw, a professional statistician. Bradshaw started his career as an actuary and moved on to (dare I say) more interesting work. Bradshaw was a sports analytics specialist for ESPN when he won this Kaggle contest. In September of 2015, he became the Manager of Basketball Analytics for the Detroit Pistons.

For Metrick's contrarian strategy and for the Kaggle arbitrage strategy, using the most predictive tournament outcome model to develop your strategy can improve your performance. But it is typically not optimal to directly submit the most predictive model to Kaggle or to submit the highest probability bracket to a bracket pool.

2.12 THE TEAM ADVANCEMENT TABLE

The team advancement table shows the probability of each team advancing beyond each round. It can be calculated from the tournament outcome model. This table is useful for estimating expected scores and other properties of the bracket.

The tournament outcome model is created before the tournament. The advancement table is calculated based on the assumption that the earlier round results do not change our assessment of team strengths and win probabilities. Instead, the approach used in tournament analysis is to make the assumption that the process of advancement is memoryless and depends only on the present state, not the previous history. This is called the *Markov assumption* because these kinds of processes were first extensively studied by the Russian mathematician Andrey Markov. There is evidence that the Markov assumption holds reasonably well for the NCAA tournament. The Sagarin ratings and the ratings of other systems are not greatly affected by games played late in the season, so one can assume they don't change much during the tournament. Also, the betting market spreads do not seem to narrow for a Cinderella team that advances deep into the tournament. The Markov assumption is also adopted for games in the same round. All tournament game outcomes are assumed to be independent events. This tournament outcome model is sometimes called a Markov model since it is used as a Markov model in tournament analysis.

Some bracket pools allow you to adjust your picks for future rounds during the tournament. In that kind of tournament, you would probably be better off using current information like betting market spreads to adjust your picks, rather than using a pre-tournament Markov model. Even if advancement in the tournament had little impact on a team's prospects, there might be injury information that would have an impact on win probabilities.

Table 2.2 shows the team advancement table calculated from the tournament outcome model shown in Table 2.1. The Round 1 advancement probabilities come directly from Table 2.1 since

TABLE 2.2   The Team Advancement
Table Based on Table 2.1

	Round 1	Round 2
Team A	.78	.455
Team B	.64	.302
Team C	.36	.147
Team D	.22	.095

we know the exact head-to-head matchups that will occur. The Round 2 probabilities are calculated based on the win probabilities against possible opponents weighted by the previous round advancement probabilities of each possible opponent. The team advancement table for an NCAA tournament would have 64 rows and six columns.

CHAPTER 3

Ratings versus Seedings

I F YOU USE A predictive rating or the betting market point spreads to pick winners, you will typically end up picking some lower seeds to beat higher seeds. Why don't the seed ranks agree with the predictive ratings? First, predictive ratings don't always agree with each other and/or the betting market point spreads. And the NCAA has never completely committed to the idea that the seeding should be as predictive at possible. The seeding is determined by a 10-member NCAA basketball committee using secret balloting. Committee members are not allowed to vote on a team when they have a family or employment affiliation with the team.

The tournament seeding does not even exactly correspond to the committee's own ranking of teams according to relative quality. The first step in the seeding process is to rank the 68 teams to produce a preliminary seed list. But, while finalizing the bracket, teams may be moved 1 or (more rarely) 2 seed levels away from the committee's team strength ranking. This is necessary to conform to certain seeding principles. For instance, teams from the same conference should not meet too early in the tournament, teams seeded higher than 5 should not face a home-crowd disadvantage, and teams should play as close to home as possible (NCAA 2016).

Before the committee votes, they are already presented with a ranking of teams based on a rating system called the Rating Percentage Index (RPI). The RPI was developed in 1980 by statistician Jim Van Valkenburg. The RPI uses only team winning percentages. Margin of victory does not figure into the RPI calculation. Almost since its inception, quantitative analysts have found flaws in the RPI. It contained no adjustment for home-court advantage, but it was revised in 2004 to include that. It puts a lot of weight on the average win percentage of a team's opponents and a team's opponents' opponents regardless of whether the team beats these opponents or not. As a result, beating a weak opponent with a weak schedule can reduce a team's RPI, and losing to a strong opponent with a strong schedule can increase a team's RPI.

Ratings that include margin of victory in their calculation can be more predictive than ratings, like the RPI, which are based solely on win percentage. But using margin of victory in the team selection and seeding process could put pressure on coaches to run up the score in games. Herein lies a dilemma. Seeding a single-elimination tournament according to relative team strength ensures that the most deserving teams with the best records are not disadvantaged by virtue of the seeding. But using the predictive ratings, the most accurate measurement of team strength, incentivizes the unsportsmanlike practice of running up the score.

The NCAA has been downplaying the importance of the RPI in the seeding process for many years. It's true that the committee members are free to use whatever information that they want to use in their decision making, but the RPI pervades the organization of the data officially used during the seeding process.

In 2018, the NCAA modified the team information sheets that are presented to the committee to add information on three ratings that use margin of victory (Vander Voort 2018). These are Jeff Sagarin's ratings, Ken Pomeroy's ratings and the ESPN Basketball Power Index (BPI). This information is added at the top of the team sheets along with the average of the three ratings. Information on two other ratings other than the RPI that are based only on win percentage (Ken Pauga's Index, the ESPN Strength of Record) were also added. The other information in the team sheet is still organized based on the RPI. (The 2018 team sheets were also modified to give away games [played on the opponent's home court] relatively more importance than home games.)

The committee already had access to all these ratings in recent years and the power to use them. "It's just being presented in a different, more convenient and more efficient format. As with any resource that the committee has at its disposal, a committee member can use it as they see fit," according to David Worlock, media coordinator for the NCAA tournament (Medcalf 2018).

The 2018 seeding seemed to be more in conformance with the more predictive ratings and the betting market predictions. The 2018 seed list ranking announced on Selection Sunday shifted toward the predictive rating systems as compared with that of the last three years (see Figure 3.1). And the first-round seeding has relatively high fidelity with the betting market point spreads, missing on only two games by only 1.5 points. But one year's data is not enough to confirm a trend with high confidence. As this book was going to press, the NCAA announced that the RPI would be replaced with a new metric in future years. So, we will never see more data to confirm this apparent trend.

If margin of victory has more influence on the tournament selection and seedings, will this lead to running up the score? The human element in the seeding process can mitigate this. The committee members have the power to adjust the seeding to compensate for the actions of any team that engages in an unseemly attempt to game a mathematical ranking system.

You can produce a pretty good bracket for your office pool without using any information other than the unfilled out bracket available on the NCAA's website and reproduced by many media outlets, assuming your office pool is a standard scoring pool with no upset incentives and top-heavy scoring (where a correct champ

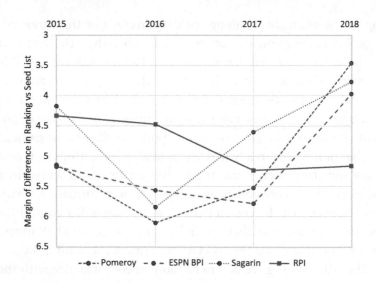

FIGURE 3.1 Difference between the NCAA seed list and four other rankings (2015–2018) showing an apparent increase in the influence of three rankings (that all utilize margin of victory) when they were added to the team sheets in 2018. (Data from Pauga 2018.)

pick is awarded a relatively large number of points). First, advance all the highest seeds in every game up to the Final Four. That is, bet no seed upsets. This will advance the four 1 seeds to be the Final Four. The strongest 1 seed (according to the committee's assessment) is in the upper left corner of the bracket by convention and the 1 seeds are ordered in strength clockwise around the bracket. Most, but not all, media sources display brackets that follow this convention. The 1 seed in the upper left will typically be over-bet in your office pool, so choose a different 1 seed to be your contrarian champ unless your pool has 10 or less entries.

> Just advancing the highest seed in every game will give you a competitive down-bracket in most office pools with standard scoring.

> Picking the 1 seed in the upper left corner of the bracket as your champ is typically a bad idea in larger bracket pools.

CHAPTER 4

The Conquest of Pools with Upset Incentives

IN THE BRACKET POOLS that Metrick analyzed, all correct picks for a given round get the same number of points. But a different kind of scoring system is used in some pools, where the players get more points for picking the underdog in a game than they do if they pick the favorite. These scoring methods call for a different strategy. Going with the favorites in the down bracket is likely to yield relatively few points versus your opponents. We also need to reconsider the strategy for picking the champ. If you can get a big point advantage over all your opponents by astutely picking upsets, then maybe it's not best to go contrarian for the champ pick.

The published strategies for pools with upset incentives use only a tournament outcome model. There is typically no source data available for estimating an opponent model prior to the deadline for submitting your brackets. These strategies have a history of working well without taking opponent behavior into account.

4.1 HOW UPSET INCENTIVES WORK

In scoring systems with upset incentives, an upset is defined by the team seeds. If a lower seed beats a higher seed, that is an upset. So, if a 9 seed beats an 8 seed then that is an upset for the purpose of assigning points. The 9 seed might be the stronger team according to the betting market spread or a rating system, but it's an upset if the 9 seed wins according to the tournament seeding.

The simplest upset incentive rule is to just award extra points. If you correctly pick a 15 seed to beat a 2 seed, then you get extra points. If you pick a 9 seed to beat an 8 seed in the first round, you get the same number of extra points as you get for a picking a 15 seed to beat a 2 seed. The number of extra points may be the same for all upsets or vary based on the round. There may be no upset points for correct picks in some rounds.

One complication is whether your bracket actually predicts an upset. Suppose you picked the 7 seed to beat the 2 seed in the second round. You predicted an upset. But, in the actual tournament, that 7 seed may end up playing and beating a 15 seed. There was no actual upset, but you correctly picked the 7 seed to win the second round. Hopefully, this detail is clarified in the pool rules. The rules can require either an actual upset, a predicted upset, or both to get the upset points.

Another common upset incentive rule is a seed multiplier rule. With a seed multiplier rule, the number of points awarded is based on the seed number of the winning team multiplied by a factor. The factor may get bigger for each round. For instance, the factors may double progressively as with the most common standard (non-upset) scoring system. For Rounds 1 through 6 the factors would be 1, 2, 4, 8, 16, 32. So, if you correctly picked an underdog 15 seed to win the first round, you would get 15 points, but if you correctly picked the favored 2 seed, then you get only 2 points. If you correctly pick the 15 seed to win in the second round, you get $2 \times 15 = 30$ points for that game. If the seed multiplier is 1, then the number of points you get is just the seed number. The seed multiplier rule tends to reward picking underdogs, but the whole issue of whether there was an actual or predicted upset can be ignored, you just get points based on the winning seed even if the favorite won.

Another upset incentive is a seed difference rule. With a seed difference rule, the number of points awarded is based on the difference between the larger seed number and the smaller seed number. The points are rewarded only if there is an upset. So, if you correctly pick a 9 seed to beat an 8 seed, you get 1 point. If you pick a 15 seed to beat a 2 seed, you get 13 points. The rules can include a multiplier factor to be applied to this difference. The issue of whether the points are awarded based on actual or predicted upsets has to be clarified in the rules.

The upset incentives are often awarded in addition to points awarded just for correctly picking the winner. For instance, you might get 2 points for correctly picking a winner of the first-round game regardless of whether it was an upset or not and also get additional points if it was an upset. Or you might get points for a correct pick plus the seed number of that pick. The number of options for the scoring rules of a bracket pool is vast.

4.2 THE EXPECTED-POINT-MAXIMIZING BRACKET

The expected-point-maximizing (EPM) bracket is the bracket, among all possible brackets, that is estimated to have the highest average (or expected) score in a bracket pool. The only strategy for pools with upset incentives that has been discussed in the research literature has the goal of maximizing the player's expected score. This is in contrast to the contrarian method of Metrick where optimization involved picking a contrarian champ in a bracket that had the highest likelihood of beating all opposing brackets.

The score that a bracket gets depends on what teams advance in the tournament. This score is not known before the tournament. The best one can do is maximize the average (or expected) score. For instance, consider the simplest possible "bracket" consisting of two teams and one game. Let's say Connecticut is playing Texas and Connecticut has a 60% chance of winning. Assume 10 points are awarded for correctly picking the winner of this game in a bracket. Then a bracket projecting Connecticut as the winner will get 10 points 60% of the time and 0 points 40% of the time. This averages out to 6 points, so the average or expected score for this bracket is 6 points. The expected score for a bracket projecting Texas is 4 points. These are the only two possible brackets, so the EPM bracket is the one projecting Connecticut. In this example, where the scoring system has no upset incentives, the EPM bracket is the one projecting the favorite to win.

The EPM bracket for a pool with no upset incentives, like the pools that Metrick analyzed, is formed simply by picking the team with the highest advancement probability among the two teams in any given game. Advancement probabilities are calculated from the tournament outcome model. Table 2.2 (in Chapter 2) shows an example of a team advancement table that includes all the required advancement probabilities for a four-team tournament. Maximizing the expected points for each game also maximizes the expected points of the bracket, since the overall bracket score is just the sum of all game points. In the 1993 bracket pools that Metrick analyzed, the EPM bracket would project no upsets and have UNC winning it all. Metrick's optimal bracket (the best bracket to play in a pool without upset incentives) would be the EPM bracket with the minimum modifications required to have an underdog like 2 seed Arizona winning it all.

But I am glossing over a detail here in equating Metrick's optimal down-bracket with the EPM down-bracket. Metrick simply assumed the higher seed in any matchup was the stronger seed with the higher win probability. But a tournament outcome model might assign the lower win probability to the higher seed in some matchups.

Metrick's approach was to just advance the highest seed in the early rounds. This can yield a point expectation that is slightly below the EPM bracket even in a pool without upset incentives. There are two reasons that this can happen:

1. There is a seeding error in the first round. For instance, the 9 seed is favored to beat the 8 seed in the betting markets. The EPM bracket would advance the 9 seed.

2. There is a seeding error that reverses the advancement probabilities relative to the seeds in later rounds. For instance, the 4 seed is slightly favored to beat the 5 seed if they meet in the second round. But, the 4 seed's first-round opponent is a good bit stronger than the 5 seed's first-round opponent, so the 5 seed is more likely to make it past the second round than is the 4 seed. In this case, the seeding-based bracket would advance the 4 seed past the second round and the EPM bracket would advance the 5 seed past the second round.

Metrick's goal was to support the contrarian champ pick by racking up as many points as possible in the earlier rounds. The EPM bracket can do a slightly better job of racking up points in the earlier rounds. So, for the purpose of implementing Metrick's contrarian strategy, it can be better to start with the EPM bracket rather than the purely seeding-based bracket.

4.3 MONTE CARLO COMPUTER SIMULATIONS

In a pool with upset incentives, determining the EPM bracket can involve some difficult trade-offs. If you analyze the first round and simply advance all the teams that provide the highest expected points, then you may find yourself advancing the 12 seed and the 13 seed. But these two teams meet in the second round, and neither may be a good option to advance to the next round, the Sweet Sixteen. You might be better off (you might get more expected points overall) if you advance the 4 seed or 5 seed in the first two rounds. We need better analytic tools to resolve all these trade-offs and determine the EPM bracket for pools with upset incentives.

David Breiter and Brad Carlin were the first to publicly tackle the problem of how to play a pool with upset incentives (Breiter and Carlin 1997). They also were the first to explicitly provide advice on how to play an NCAA tournament bracket pool in a research paper. (Metrick was just analyzing strategic behavior in markets in his paper and comparing it to a model of optimal pool betting behavior.) And, they were the first to report on using Monte Carlo methods to analyze bracket strategy.

The modern Monte Carlo method was the brainchild of the mathematician Stanislav Ulam. He was playing solitaire while in the hospital recovering from surgery in 1946 and became interested in the probability of winning. He found the problem hard to solve, but it occurred to him that it would soon be possible to solve it by repeatedly simulating the process of first shuffling the deck and then playing the game many hundreds of times on an electronic computer. Computers were just becoming capable of this feat. Ulam had worked on the Manhattan Project and immediately realized that simulations could also solve problems in neutron chain reaction research. John von Neumann soon learned of Ulam's idea, recognized its importance, and started programming the ENIAC computer to run simulations and developed supporting methods like pseudo-random number generators. The method was at first classified and needed a code name. The code name Monte Carlo was chosen based on Ulam's childhood memory of an uncle who loved to gamble there. Before Ulam's method, there had been simulation methods based on randomization using cards, spinners, dice, and even randomly dropped pins, but they had more of a supporting or confirmatory role in problem-solving. Computerized Monte Carlo simulations were soon solving problems that had no other practical solution. The method was first published in an unclassified paper in 1949.

These days it's easy to simulate a coin-flip outcome of two evenly matched teams using a computer spreadsheet. The Microsoft Excel function RAND generates a uniform random number between 0 and 1. If the value is below 0.5, then the outcome is called Heads; otherwise call it Tails. It's just as easy to simulate any game based on a tournament outcome model. If Connecticut is in a tournament game and has a 60% chance of winning, then any RAND value below 0.6 is a win for Connecticut. Chapter 2 covered how to come up with a probability matrix for the outcome of all possible games in the tournament. The complete outcome of the tournament can be simulated using the random number function and the probability matrix. The first-round matchups are already determined. The 32 first-round games must be simulated to determine the second-round matchup. Each round is simulated until the outcome of all 63 games has been simulated. This simulated outcome represents a statistical sample from the outcome space of the tournament. Using the scoring rules of the pool, you can determine the score that a particular filled-out bracket gets for this outcome. So, it's possible to build up a set of random samples from the probability distribution of scores for a particular bracket. The average of this set of scores is an estimate of the average or expected score of this particular filled-out bracket in the tournament. The larger the set of sampled scores, the more accurate the average will be, assuming the probability matrix is accurate. Hence, it's possible to use the Monte Carlo method to calculate a good estimate of the expected score of any bracket before the tournament is played.

4.4 FINDING THE EPM BRACKET

Breiter and Carlin limited their analysis to each of the four 16-team regional brackets of the tournament. They simulated a regional bracket 10,000 times to get an estimate of the average or expected score of the bracket. The error of this estimate (the "standard error") is only about one quarter of a point. The standard error calculation assumes a normal distribution of errors and indicates that the error falls within 0.25 points of the real value 67% of the time. This level of precision was more than adequate for solving the problem at hand.

They demonstrated the method for one system of pool rules that awards points for wins and also bonus points for upsets. The rules only cover the four rounds of a regional tournament. The points awarded for wins (without regard to whether they were upsets or not) were 5, 8, 15, 25 for the four rounds. The upset bonuses are based on the Seed Difference Rule. In the case of an upset, the seed difference (higher minus lower number) is multiplied by a round-specific factor. The factors are 2, 3, 4, 5 for the four rounds. For instance, if a 5 seed beat a 12 seed in the first round, then the points awarded would be $5 + (2 \times (12 - 5)) = 19$.

Breiter and Carlin analyzed four different strategies for playing a bracket.

The first strategy was just betting the highest seed, which they named "Seed Favorites." This was Metrick's presumed best strategy for a pool without upset incentives.

The second strategy, dubbed "Big Four," picked the four most likely upsets in the first round of each region. The most likely upsets were assumed to be the ones where the teams were most closely matched according to seed number. The "Big Four" strategy picked seeds 9 through 12, thereby eliminating seeds 5 through 8 in each region. These were the only upsets in the bracket sheet. Otherwise, the high seeds were advanced just as in the "Seed Favorites" strategy.

The third strategy, called "Best Trial," was a trial and error strategy that was not specified in detail in the paper. The resulting picks from this strategy turned out to be similar to those of the "Big Four" strategy. It played the first round somewhat more conservatively and the later round somewhat less conservatively than did the "Big Four" strategy.

The last strategy was called "Best of All." It amounted to a brute force search of every possible bracket. There are 32,678 possible distinct brackets that may be submitted to compete in a 16-team regional tournament. Each of these possible brackets had to be evaluated against the set of 10,000 simulated tournament outcomes in order to estimate the expected points that the bracket would glean. Overall, this took eight hours of CPU time on a Sparc20 workstation.

The "Best of All" strategy estimated the true EPM strategy. The "Best Trial" strategy uses EPM, but only within the limited range of the brackets that were tested by trial and error. The other two strategies where just fixed strategies. They were evaluated using the Monte Carlo method, but the brackets were fully specified before evaluation so there was no room for using the Monte Carlo results to improve the bracket.

The sum of the expected scores of the brackets of the four regions ranged from 316 for "Seed Favorites" to 365 for "Best of All." "Trial Best" was about 5 points behind "Best of All." The brackets fared better in the actual tournament. "Trial Best" actual score exceeded its expected score by 117 points. "Best of All" exceeded its expected score by only 68 points. As a result, "Trial Best" had a better score in the actual tournament. The standard error in the expected scores was only about 0.25 of a point. Yet, the actual scores vary from the expected scores by tens of points, or even 100 points. Why is this? The standard error quoted is only the standard error in the Monte Carlo estimate of the expected points. Other sources of error are not included. The Monte Carlo estimate assumes that the outcome model (the matrix of win probabilities for team matchups) is exactly correct, but there is some uncertainty and maybe some bias in those values. But the main source of error is the large and impactful uncertainty in the one and only actual tournament outcome.

To the extent that using a tournament outcome model is an improvement over any fixed strategy like "Big Four," EPM based on a tournament outcome model should be the way to go.

It is not practical to extend the "Best of All" strategy to the full bracket. It would take 64 billion years on a Sparc20 workstation. The calculation has to be completed in the less than four days between the announcement of the bracket seedings and the tip-off of the first game of the tournament. No current computing network could perform the calculation in time.

4.5 DIRECT CALCULATION OF THE EPM BRACKET

In June of 1999, Yale professor Ed Kaplan had a eureka moment in a hotel in Rome during a bout of insomnia brought on by jet lag. He discovered a direct calculation for the Breiter–Carlin

"Best of All" method, thereby eliminating the need for applying the Monte Carlo estimation and a brute force evaluation of all brackets. Unlike the Monte Carlo method, the direct calculation is not computation intensive. The computation completes quickly even when it is applied to the whole 64-team bracket (Kaplan and Garstka 2001).

> There is an algorithm that maximizes your expected points in pools with upset incentives.

The method is based on recursion. Recursion is a process that relies on a basic starting point and the repeated application of a calculation rule. The Fibonacci sequence is an example of an integer sequence that is created using recursion. The recursion rule is as follows: the next value in the Fibonacci sequence is the sum of the two previous values. The starting point or *base case* for the Fibonacci sequence is the sequence 0, 1. Applying the recursion rule repeatedly, the Fibonacci sequence is 0, 1, 1, 2, 3, 5, 8, 13, 21, and so on. The Fibonacci sequence is a source for one of the more popular scoring rules for bracket pools. The sequence 2, 3, 5, 8, 13, 21 is often used as round-specific points awarded for the six rounds in bracket pools. The most common scoring rule for bracket pools is the first part of another sequence that is an example of a sequence based on recursion. It starts with 1 (the base case) and each value is double the previous value: 1, 2, 4, 8, 16, 32. It's called exponential scoring because the score for each round is two raised to the power of the round number minus one. The value of a correct pick in round six is $2^5 = 32$.

As mentioned earlier, you can't maximize expected points if you start maximizing at the bottom of the bracket and go up. You may eliminate two strong seeds in the first round that would meet in the second round, thereby failing to maximize points overall. This means that a systematic approach will be top-down. Kaplan and Garstka defined the EPM bracket using a top-down recursion on the bracket.

Game outcomes are the fundamental unit of bracket scoring, so it's helpful to diagram the bracket explicitly as a binary tree linking games. Games are linked by virtue of the fact that the winning team in each game advances to another game in the next tournament round. The four-team tournament bracket shown in Figure 2.1 is converted into a binary tree linking games shown in Figure 4.1.

Figure 4.2 shows the recursive structure of the bracket. The base case is the whole bracket. In a recursion step, the whole bracket or a sub-bracket is replaced by expanding it into its crowning game and its two sub-brackets. The recursion will continue until the six tournament rounds are defined. Unlike the Fibonacci sequence, this recursion is not infinite, so it needs a terminating case, which is not shown. In the terminating case, the "sub-brackets"

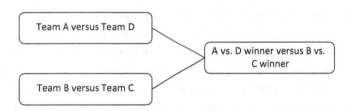

FIGURE 4.1 A four-team tournament bracket (Figure 2.1) converted into a binary tree linking games.

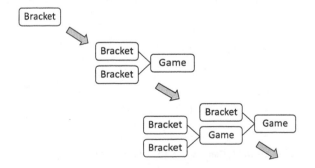

FIGURE 4.2 The recursive definition of the bracket.

representing the first round consist of a single game, so the sub-bracket symbol would be replaced by the game symbol.

Figure 4.3 shows how a submission to a bracket pool can be scored using recursion. The brackets are scored after the tournament results are in. Each correctly picked game winner is awarded points according to the scoring rules of the pool. The points from correct picks are summed to calculate the overall score. As shown in Figure 4.3 a bracket can be scored using the simple recursive rule: the bracket score is the sum of its crowning game score plus the sum of the scores of its two sub-brackets. You probably don't use a recursive rule to score your bracket. It is not necessary. But this illustrates how a top-down recursive operation can be applied to the bracket. A top-down recursion is necessary for systematically determining the bracket with the maximum score.

The average (or expected) score of any bracket can be calculated using a recursive rule: the bracket expected score is the crowning game's expected score plus the sum of the two sub-bracket's expected scores. The expected score of a game in a bracket is the advancement probability of the winner picked for that game multiplied by the score that would be awarded if that team wins. An example of the advancement probabilities for a four-team tournament are shown in Table 2.2.

The specific recursion described by Kaplan and Garstka covers a limited set of game-scoring rules that are calculated based solely on the seed numbers of the winning teams for each tournament

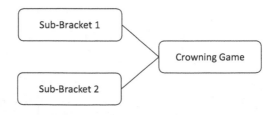

Bracket Score = Crowning Game Score + Sub-Bracket 1 Score + Sub-Bracket 2 Score

FIGURE 4.3   A recursion rule for scoring a bracket.

round, but the authors anticipated that the recursion could be applied more generally. A more general recursion can cover game-scoring rules where the seed number of the losing seed in a submitted bracket and/or the seed number of the losing seed in the actual tournament outcome bracket is a variable in the game-scoring equation. This more general recursion covers all the scoring rules described in this chapter.

In Kaplan and Garstka's recursion, the EPM bracket is the bracket that, among all possible brackets, has the maximum expected score. This EPM bracket is the bracket with the maximum expected points among the set of 64 EPM brackets that are constrained to have 64 different champions (i.e., different crowning game winners). These brackets cover all the possible outcomes. Each of these constrained EPM brackets can be defined by recursion. Figure 4.4 is a depiction of the recursive definition of the EPM bracket that is constrained to have Duke as the champ pick. The crowning game has Duke as the winner. Therefore, Duke must be the champ pick in one of the sub-brackets. Any of the 32 teams that are seeded in the other sub-bracket may be the champ of that sub-bracket. So, one of the sub-brackets must be the EPM sub-bracket constrained to have Duke as champ. The other sub-bracket must be the unconstrained EPM sub-bracket; this is the bracket with the maximum expected points among the 32 EPM brackets that are constrained to have the 32 different champions that include all the teams seeded in that sub-bracket. Both these sub-brackets are either directly defined by recursion or determined picking

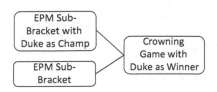

FIGURE 4.4   The recursive definition of the EPM bracket constrained to have Duke as champ.

the sub-bracket with the maximum score from a set of sub-brackets that are defined by recursion.

In the more general case, both the teams that play in the crowning game must be specified since both teams are required to determine the score awarded for that game. In this case, it is necessary to determine the bracket with the maximum expected score among the set of $64 \times 32 = 2048$ EPM brackets that cover all possible crowning game matchups. Figure 4.5 shows the recursive definition of the EPM bracket with UCLA beating Duke in the final game. The two sub-brackets used in this definition are EPM sub-brackets with the winner of the crowning game constrained to be either Duke or UCLA.

4.6 EPM FOR THE PUBLIC

In 1999, I played an office bracket pool near Duke University. This was a pool with no upset bonuses. It was the type of pool that Metrick had analyzed in his natural experiment. The scoring rule was the most common one: 1, 2, 4, 8, 16, 32 exponential scoring. The pool offered prizes for first through third. That year, Duke was a strong favorite to win the national championship. But I decided to go contrarian. I was then not aware of Metrick's paper. I bet three brackets, one for each of the three other 1 seeds: Connecticut, Michigan State, and Auburn. For the rest of all three brackets, I played the seed favorites.

After the start of the tournament, all the brackets were revealed. There were nineteen opposing brackets. Sixteen of the opposing

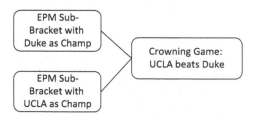

FIGURE 4.5 The recursive definition of the EPM bracket with UCLA beating Duke in the crowning game.

brackets picked Duke, one picked 1 seed Michigan State and one picked 2 seed Maryland. I was surprised by the extreme level of over-backing of Duke. One of my opponents had entered three brackets all with Duke as the champ pick.

Duke and Connecticut met in the championship game and Connecticut won. And I won the pool. My all-favorites down-bracket held up so well that I would have finished in third and still be in the money if I had bet a bracket for Duke in spite of all the competition backing that champ.

I was impressed with this strategy that involved betting all favorites except for a contrarian champion. I did some research on it, but I did not find Metrick's paper or any other good information on contrarian strategy in the bracket pool. I did find Breiter and Carlin's paper on EPM in pools with upset incentives.

Breiter and Carlin's paper was not of any practical use to me since I was not playing a pool with upset incentives, but I found the ideas interesting. Later in 1999, I had a eureka moment. The same muse that had visited Ed Kaplan in June visited me some-time in the late summer. I had also discovered the direct recursive solution for finding the EPM bracket.

I decided to make the EPM calculation available on a new bracket pool strategy website named Poologic.com along with a contrarian strategy calculator that gave results similar to Metrick's approach for pools without upset incentives. Poologic is a free website that encourages pool players to win pools using these evidence-based strategies and donate some of their win-nings to charity. Poologic was first available for the year 2000 tournament pools. I became aware of Kaplan and Garstka's paper when Stephanie Williams wrote an article for *SmartMoney* maga-zine on bracket pool strategy where she interviewed both Kaplan and me (Williams 2001).

4.7 EPM RESULTS

Carlin finished in the money in three of five years using his method in a pool with about 70 entries. (Carlin was able to take advantage of a special second-chance "insurance" rule in his pool that allowed him to insure one team for an additional 50% of the bracket submission fee. If the insured team lost a game in any of the first four rounds, then the insured team was replaced with the game winner. This rule helped him finish in the money in two of those three years.) Ed Kaplan outscored the competition in two of four years in an online bracket contest run by Professor Erik Packard of the mathematics department at Mesa Colorado University. Also, one of Kaplan's entries finished in the 99.97 percentile of the 2000 CBS SportsLine bracket contest. The entry was 25th in this large online pool with 95,000 entries. Victor Mather, a reporter and editor for the *New York Times*, using a similar system, finished in the money about one-third of the time playing a pool where less than 5% of the participants ever got a payout. I used the method in a 2017 pool and finished in the money. I finished fourth in a pool with 69 total entries. I also won money in 2018 as a result of playing the same pool. These are the only years when I played a pool with upset incentives.

Grant Wahl of *Sports Illustrated* tested the EPM method and wrote an article about it (Wahl 2005a). He submitted three brackets in his 2004 office pool using the three tournament outcome models available on Poologic.com at that time. The pool had 107 opposing entries. The brackets performed badly in his pool. One bracket had the worst score in the pool and the two others were, relatively speaking, dogs. I was shocked and enquired with Carlin and Kaplan about what went wrong. Neither thought that these results were cause for alarm. Carlin indicated that the EPM strategy was a risky one. I took him to mean that it could have very bad years and very good years because the pool score of EPM brackets tended to have a high year-over-year variance relative to the competition. The results from the next year's pool illustrated Carlin's point. Wahl would have finished in the money had he played three brackets based on the same three tournament outcome models in his 2005 office pool (Wahl 2005b).

Carlin, Mather and I used a somewhat modified EPM bracket sheet. This tended to have relatively more conservative picks in the last three rounds. Carlin did not have the EPM picks for the last two rounds available because his system did not calculate those. Also, as mentioned earlier, Carlin's pool rules had a feature that allowed him to insure against the loss of his champ pick in the first four rounds by replacing it with the team that beat it. Mather was using a heuristic of his own invention that was similar to EPM. I bet three EPM brackets that were modified so that three different 1 seeds won the championship game.

The scoring rules were different for each pool. Each pool was scored based on some combination of seed multiplier rules, seed difference rules, and rules that awarded points for all wins regardless of the seed.

The pool Mather played offered 1, 2, 3, 5, 10, and 25 points for Rounds 1 through 6 for each win. The seed difference was added to that. For instance, if a 4 seed beat a 1 seed in the third round, then the points awarded were $3 + (4 - 1) = 6$ points.

The pool Wahl played offered 0, 0, 0, 15, 20, and 25 points for each Rounds 1 through 6 for a win. The seed was multiplied by the round and added. If a 4 seed beat a 1 seed in the third round, then the points awarded were $0 + (3 \times 4) = 12$ points.

The pool I played awarded only seed multiplier points. The seed was multiplied by 1, 2, 4, 8, 16, and 32 for Rounds 1 through 6. If a 4 seed beat a 1 seed in the third round, then the points awarded were $0 + (4 \times 4) = 16$ points.

The pool Carlin played offered 5, 8, 15, and 25 points for Rounds 1 through 4 for a win. The seed difference was multiplied by 2, 3, 4, and 5 for Rounds 1 through 4 and added. If a 4 seed beat a 1 seed in the third round, then the points awarded were $15 + (4 \times 3) = 27$ points. The scoring rules for the last two rounds are not available.

Mathematician Erik Packard's free bracket contest that Kaplan played had complicated scoring factors. The rules offered 945, 1980, 3696, 6160, 9240, and 13,860 points for Rounds 1 through 6 for a win and the seed is multiplied by the same factor and added to the score. If a 4 seed beat a 1 seed in the third round, then the points awarded were 3696 + (4 × 3696) = 18,480. Packard arrived at these rules by splitting the difference between a range of opinions about how a pool should be scored. "Some people think each game in a round should count the same. Some people think you should get 2 times as many points for picking a 2 seed versus a 1 seed, 3 times as many for a 3 seed, and so on. My scoring system goes 1/2 way between. You get, for example, not 8 times as many points for picking an 8 seed compared to picking a 1 seed and not 1 times as many, but 1/2 way between, that is (1 + 8)/2 = 4.5 times as many points. The scoring system also makes any round worth 1.5 times as much as the next round in the following sense: If you assume all the better seeded teams win you will find that the sum of points in any round is 1.5 times as much as the next round. This is the contest that encourages you to pick that 12 seed to win that first round game, and that 5 seed to make it to the final 4" (Packard 2001).

These results give a positive impression of the return on investment from using an EPM bracket. But the results can be viewed skeptically as just a collection of anecdotes. It is not a statistical estimate or statistical proof of EPM's success. It may not even be an unbiased compilation of field data, since it includes only overall success stories with no reliable estimate of the number of failure stories.

> The available evidence seems to indicate that maximizing your expected points is a powerful strategy in pools with upset incentives.

4.8 STRENGTHS AND WEAKNESSES OF THE EPM BRACKET

The EPM bracket tends to pile up a lot of points against the competition in the first two or three rounds of the tournament. But the picks could be too risky for the last three rounds for some scoring rules.

The EPM bracket is only based on a tournament outcome model. It does not take into account opponent play or the size of the pool. Opponent strategy and pool size figured into the optimal strategy that Metrick calculated for pools without upset incentives. There is typically no source of data available before the pool on how bettors will make their picks in a pool with upset incentives. Methods that use both an opponent model and an outcome model can maximize your win probability, not just your expected score.

The EPM calculation can pick a relatively unlikely champ. Theoretically, if a pool awarded a very large number of points (like, for instance, one billion points) for a 16 seed to become champ while awarding few points for other seeds, EPM calculation could pick a 16 seed for champ, assuming the opponent model gave the 16 seed a small chance of winning against any opponent. This "unlikely champ" effect may be strong enough to matter for seed-based scoring with exponential multipliers (1, 2, 4, 8, 16, 32) per round when no points are awarded for wins that are not upsets. The EPM calculation picks 4 seed West Virginia as champ when the tournament outcome model described in Chapter 2 was used at Poologic in 2017 (see Figure 4.6). If West Virginia wins, then that adds 32 × 4 = 128 points to the score. The model gave West Virginia a 7.6% chance of winning the tournament, so the expected points for a win were 128 × .076 = 9.7 points. The strongest 1 seed,

FIGURE 4.6 The 2017 EPM bracket for the tournament outcome model described in Chapter 2 for the exponential seed-multiplier scoring rule.

North Carolina, had a 13.9% chance of winning, so the expected points for North Carolina were 32 × 0.139 = 4.48 points. The EPM calculation chose West Virginia as the champ because it was the team that overall provided the highest expected point total.

In our bracket pool play, Carlin, Mather, and I substituted some less risky teams in the last three rounds for those selected for the EPM bracket. But it is not clear if that is always a good idea. Kaplan's results and Wahl's results and findings indicate that pure EPM brackets can perform well. The EPM bracket often wins in a pool with upset incentives even if it does not correctly pick the champ.

Among the scoring rules that I have observed in actual office pools, the exponential seed-multiplier scoring rule with no other points awarded is the scoring rule most prone to picking unlikely champs. This scoring system has interesting effects on the EPM bracket. Depending on the year and the outcome model used, the Final Four may include a 1 seed or a 10 seed. Some years the highest (best) seed in the Final Four is a 4 seed. Depending on the year, the champ pick may be very conservative or more of a long-shot.

There is a wide variation in pool scoring rules. Scoring rules are usually top-heavy: a relatively large number of points are awarded for picking the overall champion. This reduces the chance that the pool winner will be decided before the championship game is played. If it is almost always necessary for an EPM bracket to pick the correct champ to win, then betting a more likely champ might be warranted. In some pools, the upset bonuses may be very small so that the pool is more like a pool with no upset incentives where a contrarian champ pick is warranted. Given the variety of scoring rules, there can be no single overall strategy for when or if the later rounds of an EPM bracket should be modified.

Also, the size of the pool matters. In a large pool, if you modify an EPM bracket to have a champ that is popular in opposing brackets, then you will have more competition if that popular team wins the tournament.

4.9 SURPRISING REACTIONS TO EXPECTED-POINT MAXIMIZERS

The high success rate of EPM brackets in pools with upset incentives has proved to be disruptive in some bracket pools.

Consider the case of Victor Mather (Mather 2003, 2011). Beginning in 1996, Mather used a self-styled system to figure out how to maximize his score, "I would take the Sagarin rankings and a calculator and figure which teams were 'good bets' depending on the bonus point system in the pool." Mather came up with a way to estimate an EPM bracket on his own, without the benefit of any of the published papers.

It's not unusual to find that some bettors already know (or assume they know) the gist of what academic researchers discover and publish about favorable strategies in gambling games. As mentioned earlier, Metrick discovered evidence that bracket pool players were already using a contrarian strategy in bracket pools. Ed Thorpe published the first research paper confirming a favorable blackjack strategy based on card counting (Thorpe 1961), but when he arrived at the Nevada casinos to give card-counting a field test, he found that there was already a lore dating back many years about card counters at the casino blackjack tables (Thorpe 1966).

Mather was enough of a statistician to figure out the basics of EPM on his own. Mather apparently knew how to use the normal distribution to convert a point spread into an estimate of the chances of winning. Mather describes his process as follows: "For each team, I multiply its chance of winning the game by the points I would get by correctly picking it in the pool. The result is the 'expected value' of each pick." The Kaplan–Garstka algorithm does exactly the same thing for the first round. Mather strategically avoided advancing too many upsets. "Now, not quite every game is that simple because you don't want to knock too many favorites out. By picking [1996 13 seed] Princeton, I was forgoing potential second- or third-round points should U.C.L.A. keep winning. One way to solve this is to pick only one upset per pair of games. If you pick the 13 to beat the 4, don't also pick the 12 over the 5. (You don't usually want to be stuck with a 12 against a 13 in the second round.)" (Mather 2011). The Kaplan–Garstka algorithm does a better job of analyzing these tradeoffs. Also, the exact calculation of the advancement probability of a team is more complicated beyond the first round, because it's not just based on the teams that are advanced from the previous round. Overall, Mather's system is not as thorough and accurate as the Kaplan–Garstka algorithm, but it's hard to know how much difference that would make for the scoring rules of Mather's pool.

Mather advises abandoning the EPM approach at the Final Four. "When it comes to picking your Final Four and winner in such a pool, it's best to take favorites. If all has gone right, you'll have racked up so many early-round bonus points that you won't need some off-the-wall Final Four choice to win." This may be good advice for some scoring systems, particularly if all the favorites would be eliminated in the Final Four on an EPM sheet for a particular tournament outcome model.

Over six years of play in a pool with 250 entrants, Mather had a first, a third, and a fifth place. That would be a highly unlikely winning streak for a player of average skill. After this streak, the pool organizer sent him an email accusing him of "cheating by picking too many longshots." Of course, all Mather was doing was playing a favorable strategy that was made possible by the very pool rules that this pool organizer had chosen to use. The next year, the pool invitation contained the following note: "The Board has discovered that some entrants in previous years have used a formula for trying to pick upsets. This runs counter to the spirit of the pool and the awarding of bonus points. The purpose of the pool is to pick winners of games and be rewarded by identifying unlikely victors. It is not to capitalize on a formula and gain an edge on entrants who are playing by the spirit of the rules. Therefore, any entry deemed by the Board to be a 'formula play' will be rejected."

Mind you, the Board could have just changed the pool rules. But instead of doing that, they outlawed good bracket pool play! The organizers had always fared well in this pool, so maybe they were just trying to keep Mather from cutting into *their* advantage.

Another case involves the wife of economist Craig Newmark (Newmark 2003). Only, this case involves expected point *minimization*.

There was an office pool at Newmark's wife's workplace and she wanted to play. But she did not follow college basketball, so she asked for assistance. Newmark, being an economist, immediately asked how much money they were going to win. She told him half the money goes to the top scorer. And there was a consolation prize for the player that finished last: the other half of the pot.

Modest consolation prizes are not that unusual. The Tour de France awards the *Lanterne Rouge* (Red Lantern) to the cyclist that finishes last. A red lantern may be hung at the end of a train as a warning to oncoming trains. Cyclists actually compete for the *Lanterne Rouge*, and it takes some strategy to finish last since you get completely eliminated from the tour if you finish too far behind on any given day. And, a number of bracket pools have consolation prizes. But consolation prizes tend to be small, like return of the entry fee. Half the pot is a really big consolation prize.

Newmark prepared two brackets. One bracket was meant to achieve a high score. The other bracket was designed to win the consolation prize. In the bracket aimed at winning the consolation prize, the lowest seeds were advanced. The 16 seeds (none of which have ever even won a first-round game at the time) all made it to the Final Four. That meant that two 16 seeds had to compete in the championship game and one had to win.

After the first two rounds of the tournament, Newmark's wife was firmly in last place, having scored only 7 points. The next worst bracket had 29 points. Every one of her Sweet Sixteen winners had been eliminated, so she could score no more. She was already locked in to win half the pot. Her coworkers had some consoling words for her, but she told them that she had done it deliberately. She explained that "half the pot was a nice amount and that economists predict that people respond to incentives."

Her coworkers were shocked. She was accused of "impropriety" and banned from the bracket pool for one year. In this case, the pool organizers also employed the obvious remedy of changing the pool rules.

Craig Newmark summed up the situation pretty well: "The moral of the story is that thinking like an economist may make you unpopular but can make you some money."

CHAPTER 5

Predicting Your Opponent's Brackets

I N 2000, MICHIGAN STATE was awarded a 1 seed. But they had seven losses, the most losses ever for a 1 seed. They were not at the very top of the polls or the ratings. They did not look to be the strongest team in college basketball. Back in 2000, the NCAA selection committee did not formally rank all teams in order of strength during the seeding process. So, the seeding provided no information about the relative rank of the four 1 seeds. These days, the strongest 1 seed (according to the selection committee assessment) is located in the upper left corner of the official bracket and the order of strength in the bracket proceeds clockwise around the bracket, with the weakest 1 seed found in the lower left corner. Michigan State was seeded in that lower left corner, but this meant nothing in the year 2000.

Even though Metrick's 1996 paper had received little attention by 2000, some pool players had surmised that they could improve their bracket by identifying and avoiding a bet for a champion that was the most over-backed team in their office pool. But, back in 2000, they had no good way to determine the most over-backed team.

Your opponent's brackets start rolling in to the pool manager after the seeding is announced on Selection Sunday. But most pool managers keep the pool entries secret until after submissions are closed around noon on Thursday. Data on opponent picks exists, but it is not made public.

When the pool manager revealed everyone's picks in my office's pool in 2000, I was surprised to see that Michigan State was the most over-bet team in the pool. Even with this information in hand, it is not clear why it happened. Perhaps it was because Michigan State's star point guard, Mateen Cleaves, was 2000s chosen media darling. This charismatic three-time all-American received the lion's share of the media coverage before the tournament began.

By the time the 2002 tournament rolled around, this guessing game ended. Yahoo and ESPN started revealing the pick distributions of their online nationwide bracket contests (Yahoo 2018; ESPN 2018). They make the early pick distribution available as early as Monday afternoon after Selection Sunday, more than two days before the deadline for submitting brackets to office pools. ESPN calls their pick distribution "Who Picked Whom." CBS also provides pre-tournament data on the pick distribution of their national bracket contest in articles on their "Bracket Challenge" website. Yahoo simply calls theirs the "Pick Distribution."

> The pick distributions from large nationwide bracket contests can be used to estimate the behavior of your office pool opponents.

In this chapter, we are going to discuss how to use a pre-tournament pick distribution to build an *opponent model*. The opponent model can be used to estimate the set of opposing brackets in your office pool. These are the brackets that you must outscore in order to win your pool.

5.1 DATA SOURCES FOR AN OPPONENT MODEL

The Yahoo and ESPN pick distributions may be the only reliably available sources of data for building an opponent model. It might be possible to locate the CBS pick distribution, and some other smaller online pools may provide their pick distributions in advance of the start of the tournament.

Compared to the tournament outcome model, the data sources for building an opponent model are limited. There are dozens of rating systems that can be used to create a tournament outcome model. Over 400 teams built their own tournament outcome models to compete in the Kaggle March Machine Learning Mania 2017 competition. If you did not want to use an existing rating system, you could build your own from the data available on each team's pre-tournament performance and (perhaps) other team characteristics.

The pick distributions take the form of a pick advancement table with 64 rows and six columns. The pick distribution is compiled from all brackets submitted to the bracket pool or contest. There is one row for each team and one column for each of the six tournament rounds. Each round's column shows the percentage of submitted brackets that advanced the team represented by the row. These advancement percentages may be converted to advancement proportions that represent estimates of the advancement probabilities simply by dividing the percentage by 100.

5.2 ADVANCEMENT TABLE BIAS

Is the pick distribution of these large nationwide bracket contests an accurate representation of the pick distribution of your office pool? It will be reasonably accurate if the two distributions are sampled from the same population of submitted brackets. Even if the two samples are from the same population, then there will be some differences in the distribution just due to chance. The 2015 ESPN pick distribution indicates that 48.8% of brackets picked Kentucky as champ. But a 2015 office pool with 100 entries drawn from the same population will not typically have exactly 49 entries that pick Kentucky as champ. There will be some random variation around the expected value of 49%.

Some skepticism is warranted about the hypothesis that your office pool brackets are drawn from the same population as those submitted to a nationwide bracket contest. Metrick discovered that there is a hometown effect where people over-bet the local favorites. The hometown effect will introduce a bias in your local pool's pick distribution relative to that of the nationwide contest if a local team is competing in the tournament. And there are other differences that might be important. Metrick discovered that there was more contrarian betting (i.e., betting an underdog) for champ in larger pools. The pool scoring rules in your pool may be different from those of the nationwide contests. You may have other information on the preferences of specific competitors in your office pool. For instance, you might know a competitor's alma mater or you might know (from their past brackets) that they always favor Duke in their picks.

Brad Null (2016) provides some measurement relevant to the hometown effect. Null called it "homer bias." He did his analysis at the granularity of a state. The people submitting the brackets were classified by the state where they resided, so he specifically measured the home state effect. The data he used came from the CBS Bracket Manager. The bias was stronger in later tournament rounds and stronger for longshots than for favorites. Relative to the nationwide pick rates, favorites (teams with greater than a 5% chance of winning the tournament) were 38% more likely to be picked to be one of the four regional winners and 159% more likely to be picked to win the tournament. For instance, if a specific favorite was picked as champ by 10% of the brackets submitted to the CBS contest, then homer bias bumps that up to 25.9% on average. Long shots (teams with less than 1 chance in 1000 of winning the tournament) were 1500% more likely to be picked for champ in a home state bracket. Some teams also showed a regional bias that extended beyond the boundaries of their home state.

The Yahoo and ESPN contest have standard scoring rules with no upset bonus (i.e., no bonus points for picking lower seeds). So, the pick distributions are not considered to be good representations of the pick distributions of pools with upset incentives. There is no published research on estimating opposing picks in bracket pools with upset incentives.

Some pool managers make the current pick distributions public while bracket submissions are still open. Some pool managers even reveal the submitted bracket as they are submitted. But revealing the brackets before the tournament starts is a rare practice and there is no published research on how to best take advantage of the information revealed due to this practice.

5.3 A PICK ADVANCEMENT TABLE EXAMPLE

Table 5.1 is an example of a bracket pick advancement table for a four-team tournament like the one depicted in Figure 2.1.

The table is similar to the tournament advancement table shown in Table 2.2. But there is a difference. The tournament advancement table was derived from a probability model called the tournament outcome model. So, while the tournament advancement table is a table of probabilities, this pick advancement table is a compilation of raw data. The values represent proportions of actual picks in a tournament. If the table was based on a bracket contest with 1,000,000 submissions, then 0.61 represents the around 610,000 brackets that picked Team A to win the tournament. The value 0.61 is the proportion of brackets that picked

TABLE 5.1   A Pick Advancement Table

	Round 1	Round 2
Team A	.87	.61
Team B	.69	.28
Team C	.31	.06
Team D	.13	.05

Team A for champ rounded off to two digits. In order to make rounded proportions have the characteristics of probabilities, the proportions might have to be normalized so that the sum for all the champ picks equals 1.0.

We will treat these advancement proportions as accurate estimates of the advancement probabilities. But in some situations, there may be a good reason to assume that there is some sampling error in this raw data.

5.4 SOME LAWS OF PROBABILITY

A few laws of mathematical probability are useful for understanding advancement tables.

Probability models have a sample space consisting of all possible mutually exclusive outcomes. The sample space of a coin flip contains two outcomes, heads or tails. The sample space of the outcome of a roll of a single six-sided die is the possible number of dots on the upward face when the die comes to rest, one through six. The probability of the complete set of outcomes must add up exactly to one (1.0).

The probability of combinations (or unions) of the mutually exclusive outcomes from the same sample space is equal to the sum of the probability of each outcome. If the probability of getting a one is 1/6 and the probability of getting a two is 1/6 in a dice roll, then the probability of getting either a one or a two is $1/6 + 1/6 = 2/6$.

The probability of joint outcomes is the product of the outcomes. If the probability of getting a head on a coin flip is 1/2, then the probability of getting two heads on both of two flips of a coin is $1/2 \times 1/2 = 1/4$. But this is true only if one outcome in no way affects the other outcome. If the outcomes have this quality, then they are said to be *independent*.

If you glued two coins together (as depicted in Figure 5.1) and then flipped them, the outcomes would not be independent. The outcome for coin flips would still be random and the probability

FIGURE 5.1   Two coins glued together.

of getting a head could still be 1/2 for each coin. You'd just have to flip the two coins together as a single unit. If the probability of getting a head on one of the coins is 1/2, then the joint probability of getting heads on both would be 1/2 rather than 1/4 and the outcomes would be positively correlated. If the coins were glued together face to face, then the probability of getting heads on both would be 0 rather than 1/4 and the outcomes would be negatively correlated.

The basic mathematical laws of probability were first discovered by Gerolamo Cardano. Cardano was a professor of mathematics, physician, inventor, and gambler. Born in Italy in 1501, he is one of the more colorful characters of the Renaissance. He invented the universal joint (the Cardan shaft). He traveled to Scotland to cure a severe respiratory disease of the Archbishop of St. Andrews that could not be cured by the efforts of the most accomplished doctors of northern Europe. He told the Archbishop to get rid of his feather pillow.

Cardano wrote *The Book on Games of Chance*, the first book that presented a mathematical theory of probability. He called the sample space of possible outcomes the "circuit." He introduced a principle of equality as the ideal for a gambling apparatus (and other aspects of gambling). An ideal die was equally likely to fall on any of its six sides. These days this is called the *principle of indifference*. Since the six sides of the ideal die have equal probabilities and must sum to 1.0, it follows that the probability of each side resting up after a die roll must be 1/6. Cardano also hypothesized the Law of Large Numbers, that the outcome frequencies of a large number of rolls of a fair die would tend toward the ideal value of 1/6.

After several failed attempts (the failed attempts are still in his book), Cardano arrived at the mathematical rule that the joint probability of independent outcomes is the product of the probabilities of the outcomes. He did not prove the rule, he just found that it worked on certain examples. He did not discover the concept of independent outcomes. All the outcomes he worked with where either mutually exclusive or independent.

Cardano discovered the concept of a probability distribution. The first distribution that Cardano analyzed was that of the six-sided die. This probability distribution is a special case of what is now called a *categorical distribution*. A categorical distribution is defined by a discrete number of outcomes (categories) and a probability for each outcome. The categorical distribution for the rolls of a single die has six categories and each category has a probability of 1/6. Cardano only worked with categorical distributions with equiprobable categories or other categorical distributions that could be derived from them via combinatorics. He derived and worked with categorical distributions for multiple dice and random samples of multiple cards from a card deck. He did not derive the more general categorical distributions that can be used to model loaded dice rolls or biased coin flips or the outcomes of basketball games or tournaments among teams of unequal strengths.

Cardano never published his book on probability. It remained more of a working manuscript throughout his lifetime and still contained a number of errors at his death. It was published 100 years later. By the time the book was published, the principles of mathematical probability had been discovered by others.

5.5 PICK ADVANCEMENT TABLE CHARACTERISTICS

The advancement table must conform to the mathematical laws of probability. Each tournament game has a sample space and all the outcomes in the sample space must add to 1.0. We can enumerate the list of teams that can be picked to meet in any tournament game. We can do this before the tournament based on the structure of the bracket. Two specific teams can meet in each first-round game, any of four specific teams can participate in each second-round game. The number doubles for each round. Any of the 64 teams in the tournament can, in principle, participate in the last-round championship game. The set of teams that can meet in a game defines the sample space of possible picks for the game.

In our example pick advancement table (Table 5.1), Team A will play Team D in the first round, so their first-round advancement probabilities must add to 1.0. Team B will play Team C in the first round, so their first-round advancement probabilities must also add to 1.0. Any of the four teams could win in the first round and participate in the second-round game, so the sample space for the winner of the second-round consists of all four teams. Therefore, the second-round advancement probabilities of all four teams must add to 1.0.

If a team is picked to play in a particular game, it must be picked to win or lose that game. This defines another sample space. If Team A is picked to play in the second-round game, then it must be picked to win or lose the second-round game. Team A has an 87% chance of being picked to play in the second-round game and a 61% chance of being picked to win that game, so it has an 87–61 = 26% chance of being picked to lose that game. Putting it another way, reasoning from a concrete example, if there are 100 brackets submitted to a pool and 87 pick Team A to win in the first round and 61 pick Team A to win in the second round, then, of necessity, 26 of the brackets in the pool must pick Team A to lose in the second round. Hence, starting with the pick advancement table, we can calculate the probability of any team being picked to lose in any round.

The tournament outcome model developed in Chapter 2 is assumed to be a Markov model that applies to all tournament games. The outcomes of the first-round games are assumed to be independent events.

If we assume that the first-round picks in a pick advancement table are independent, then this puts certain constraints on the pick advancement table. The event of Team C being picked to win its first-round game is independent of the event of Team D being picked to win its first-round game. These picks are like two independent coin flips of two different coins. Based on the Markov assumption and the rule that the joint probability of independent events equals the product of the probability of each event, we can calculate the probability that Team C is picked to play Team D in the second round is $0.31 \times 0.13 = 0.0403$, a little more than 4%.

When Team C and Team D are picked to play each other in the second round of a bracket, it is necessary to advance one of them to win the second round. So, the sum of the second-round advancement probabilities of Team C and Team D must add to a value exceeding 4% in the pick advancement table. They sum to 11% in Table 5.1, so the sum meets this constraint.

However, Table 5.2 violates this constraint.* The advancement probabilities of Team C and Team D sum to only 4%. Is this a valid pick advancement table or is there something wrong with it? This table violates the Markov assumption. The advancement picks of the two first-round game winners cannot be independent. The advancement probabilities in Table 5.2 imply that picking Team C to win the first round must be negatively correlated with picking Team D to win the first round.

The Markov assumption that the actual first-round *game outcomes* (as opposed to the *bracket pick* guesses made by people well before the game is played) are independent seems pretty compelling. Two first-round games may be played at the same time. In

---

* Mason Wright, a graduate student at the University of Michigan, discovered this in unpublished research.

TABLE 5.2   A Pick Advancement Table
That Is Inconsistent with a Markov Model

	Round 1	Round 2
Team A	.87	.65
Team B	.69	.31
Team C	.31	.03
Team D	.13	.01

that circumstance, it would be hard to see how the outcome of one game could affect the outcome of another. Even when one is played after another, it's hard to see how the outcome of an earlier game could have much of an effect on the outcome of a later game. But, is the Markov assumption really compelling for *bracket picks*?

It's more plausible that first-round *picks* might be dependent. For instance, Victor Mather recommended that both the 12 seed and 13 seed should not both be picked to advance in the same region because that forces the selection of an unlikely pick to advance to the Sweet Sixteen (Mather 2011). If many people were to adopt that strategy, then this could lead to a negative correlation between picking the 12 seed and 13 seed to advance in the first round.

The Markov assumption is often useful as a simplifying assumption. But in this particular application to the opponent model, it could be a bit of a complicating assumption since some possible advancement tables cannot be generated by a Markov model.

5.6 CONVERTING AN ADVANCEMENT TABLE INTO AN OPPONENT MODEL

In Chapter 2, the tournament outcome model, a Markov model, was created first and then a tournament advancement table was calculated from the outcome model. But the pick distributions available before the tournament begins provide only the pick advancement table. Working with an advancement table to create a model of opponent behavior is one of the least well-researched aspects of the bracket improvement process.

5.6.1 The mRchmadness Method

Eli Shayer and Scott Powers have developed a simulation-based method for estimating the bracket that is most likely to win a bracket pool (Shayer and Powers 2017). The method is freely available as an R package entitled mRchmadness. The package simulates opponent picks based on an advancement table. R is an open source computer language and environment for statistical computing (r-project.org).

The mRchmadness method does not attempt to convert the advancement table to a Markov model. The simulation resulting from the mRchmadness method is not necessarily consistent with a Markov model. The results may violate some of the conditions of a Markov model. This is not necessarily a bad thing since there is no strong reason to assume that the human process of making bracket picks is a Markov process.

The overall goal is to simulate opponent brackets using a simulation algorithm that is consistent with all the pick probabilities in the pick advancement table. For instance, Table 5.1 has Team A picked for champ 61% of the time and picked to win the first round 87% of the time. So, a simulation based on Table 5.1 must pick Team A for champ 61% of the time and pick Team A to win its first-round game 87% of the time. Similarly, the goal of the opponent bracket simulation is that all pick simulations for all rounds should conform to the probabilities in the pick advancement table.

The mRchmadness method starts by first picking a winner for the championship game and then working backward to eventually fill out the entire bracket. The mRchmadness package automatically loads the advancement table from the ESPN Bracket Challenge. But the user may create and use a different pick advancement table. The advancement table provides estimates of the pick advancement probabilities for each team in each round.

So, we have an advancement probability for each of the 64 possible champs. The probabilities add up to a total of one since they represent the universe of all possible outcomes for the champ pick.

This probability distribution is a categorical distribution with 64 categories.

The mRchmadness method predicts the championship pick by simulating a random pick from the categorical distribution that is based on the pick probabilities in the advancement table column for round six, the championship round, of the tournament. This one simulation actually determines the picks for six games, because the champ pick must also be picked as the winner of all of its five games in earlier rounds.

Once the winner of any game is picked, it is possible to derive a categorical distribution for the losing pick for that game. A categorical distribution for the losing pick may be derived by assuming that the losing pick in a game is not conditional on the winning pick. For instance, it is assumed that the probability that a 4 seed is picked to lose a Sweet Sixteen game is not influenced by the fact that a 1 seed is picked to win that same Sweet Sixteen game. This is a simplifying assumption that allows the derivation of a unique categorical distribution for the losing pick from the information available in the pick advancement table and the bracket structure. This assumption that the losing pick is independent of the winning pick is not generally consistent with a Markov process simulation where the games would be simulated based on head-to-head team probabilities that would typically be different for different possible head-to-head matchups.

Determining (by simulation) the loser of any game beyond the first round automatically determines the winner of at least one additional game in an earlier round. And the loser of any game can be determined once the winner of the game is determined. Therefore, the mRchmadness algorithm can recursively simulate the entire bracket using a collection of categorical distributions that are derived from the advancement table. And, the categorical distributions are derived in a manner that guarantees that the simulation will be consistent with the original pick advancement table probabilities.

5.6.1.1 Using the Categorical Distributions

Your opponent's behavior in the game rock-paper-scissors can be modeled as a categorical distribution with three categories since there are three possible outcomes. For instance, you might assume that each of the three outcomes is equally likely. So, you would assign a probability of 1/3 to each outcome so that the three outcome probabilities add up to one. Or you might assume that your opponent had different pick probabilities based on historical pick patterns of the particular opponent or of players in general.

A coin flip is a special case of the categorical distribution that is called the Bernoulli distribution. A coin flip has only two possible outcomes, but there are 64 possible outcomes for the champ pick in the bracket filled out by one of your opponents.

Table 5.1 is an example of a bracket pick advancement table for a four-team tournament. The probability that a team is picked to lose in Round 2 is equal to the probability that it is picked to win Round 1 minus the probability that it is picked to win Round 2. These probabilities are presented in Table 5.3 in a third column inserted to the right of the columns of Table 5.1.

When Team A or D is picked to win Round 2, then the loser must be Team B or C. The mRchmadness method assumes that the losing pick is independent of the winning pick. So, the B:C ratio of losing picks must be 41:25 regardless of whether Team A or Team D is the winning pick. We have to normalize this ratio so that the proportion of A and D losing picks add to 1.0 while keeping the ratio. The normalization is accomplished by dividing by $41 + 25 = 66$.

TABLE 5.3   Round 2 Losing Pick Probabilities

	Round 1	Round 2	Probability of Losing in Round 2
Team A	0.87	0.61	$0.87 - 0.61 = 0.26$
Team B	0.69	0.28	$0.69 - 0.28 = 0.41$
Team C	0.31	0.06	$0.31 - 0.06 = 0.25$
Team D	0.13	0.05	$0.13 - 0.05 = 0.08$

When Team B or C is picked to win Round 2, then the loser must be Team A or D. The A:D ratio of losing picks must be 26:8. We have to normalize this ratio so that the proportion of A and D losing picks add to 1.0 while keeping the ratio. The normalization is accomplished by dividing by 26 + 8 = 34.

So, now we have all the probabilities required to simulate this small tournament. First, we simulate the winning pick based on the categorical distribution represented by the Round 2 column in Table 5.1. If A or D wins, then we simulate the loser based on the categorical distribution represented in Table 5.4. If B or C is picked, we simulate the loser based on the categorical distribution in Table 5.5. Of course, the categorical distributions in Tables 5.4 and 5.5 have just two categories. Simulating them amounts to simulating the flip of a biased coin. In this small tournament, all we need to do is simulate the winner and loser of the championship game. These two teams must be the winners of the two first-round games, and these are the only first-round games in this small tournament.

We can now show that this simulation is not consistent with a Markov model. The Markov assumption is that the first-round picks are independent. So, picking Team A is independent of

TABLE 5.4   Losing Team Pick Proportions When Team A or D Wins Round 2

	Probability of Losing in Round 2 When A or D Is Picked for Champ
Team B	41/66 = 0.621
Team C	25/66 = 0.379
Total	1.000

TABLE 5.5   Losing Team Pick Proportions When Team A or D Wins Round 2

	Probability of Losing in Round 2 When B or C Is Picked for Champ
Team A	26/34 = 0.765
Team D	8/34 = 0.235
Total	1.000

picking B. Under this assumption, the probability that A and B will be picked to meet in the second round is .87 × .69 = 0.6003. If A meets B in the second round, then either A beats B or B beats A. In the simulation, the probability that A beats B is 0.61 × 0.621 = 0.379 and the probability that B beats A is 0.28 × 0.765 = 0.214. The sum is 0.593. Since 0.6003 is not equal to 0.593 the A pick is not quite independent of the B pick. So, the mRchmadness simulation is not simulating a Markov model.

5.6.1.2 Simulating Categorical Distributions

An outcome from a categorical distribution like a dice roll can be simulated in Excel using the RAND function. The RAND function returns a uniform random number between zero and one. The random number is equally likely to be in the interval between 0 and 1/6 as it to be in the interval between 1/6 and 2/6. So, if the random number is less than 1/6, then that is a roll of a one. If it's greater than 1/6th and less than 2/6th, then that represents a die roll of a two, and so on, up to the interval above 5/6, which represents a die roll of a six.

Simulating your opponent's championship pick involves a categorical distribution with 64 possible outcomes, since there are 64 teams that may be picked to win the tournament. The 64 pick probabilities are represented by the last, or rightmost, column in the advancement table. These pick probabilities divide the interval between zero and one into 64 distinct unequal intervals, just as the probabilities for the six outcomes of a die roll divide the interval between zero and one into six distinct equal intervals. Just as with the die roll, we can simulate an opponent's champ pick by running the RAND function to get a random value between zero and one and then determine in which of the 64 intervals this random value falls.

5.6.2 Reverse Engineering a Markov Model

There is a method for estimating Markov models for game picks from a pick advancement table of pick probabilities that were calculated from a Markov model. The method involves solving simultaneous linear equations (also called systems of linear equations or linear systems).

Simultaneous linear equations do not always have one solution. A set of simultaneous linear equations may have zero solutions, one solution, or an infinite number of solutions. For instance, these two equations in two unknowns have one solution for the price of an apple and the price of an orange:

2 apples + 3 oranges cost $5

1 apple + 1 orange cost $2

One way to solve systems of simultaneous linear equations is to plot the equations and see where the lines intersect. Figure 5.2 shows that the only solution for these two equations is that apples and oranges both cost $1.00.

These two simultaneous equations have no solution:

2 apples + 2 oranges cost $2

1 apple + 1 orange cost $2

FIGURE 5.2 Simultaneous linear equations with one solution.

Figure 5.3 shows that the lines are parallel and never intersect.

And these two equations have multiple solutions:

2 apples + 2 oranges cost $4

1 apple + 1 orange cost $2

Figure 5.4 shows that the lines are parallel and fall right on top of each other, so this system of simultaneous linear equations has an infinite number of solutions.

If you start with a Markov model and convert it to an advancement table and then try to convert it back to a Markov model, then you will get back your original Markov model only if the set of simultaneous linear equations has only one solution. If there is more than one solution to the simultaneous linear equations, then you would need to add some other constraints to arrive at one solution.

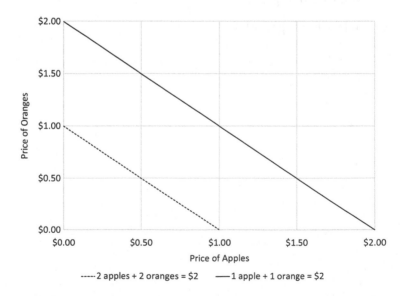

FIGURE 5.3 Simultaneous linear equations with no solution.

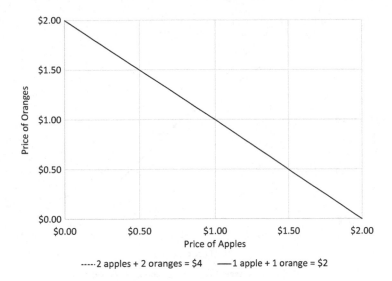

FIGURE 5.4   Simultaneous linear equations with many solutions.

It's also possible for simultaneous linear equations to have an unrealistic or impossible solution. Figure 5.5 shows the solution for the following system of simultaneous linear equations:

3 apples + 2 oranges cost $3

1 apple + 1 orange cost $2

The lines intersect at the point where an apple is worth less than zero. An apple is worth –$1. This solution makes no sense or at least would require a very odd interpretation. The merchant appears to be paying you to take his apples.

A probability value cannot be less than zero or greater than one. So, if the solutions to a system of simultaneous linear equations are supposed to be probabilities, then none of the solutions can be less than zero or greater than one. This may occur because of incorrect assumptions. Or, it may also occur due to some error or uncertainty in the numerical parameters (like an error in the costs data for apples and oranges).

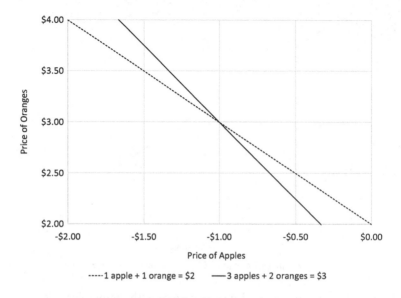

FIGURE 5.5 Simultaneous linear equations that have an unrealistic solution. The price of an apple is a negative number.

It may not be possible to reverse engineer a pick advancement table into a Markov model. It's always possible to derive sets of simultaneous linear equations from an advancement table, but some of the sets of equations may have no solution or an impossible solution. The advancement table may not have been generated by a Markov process. Or the advancement table might not have full fidelity to the Markov process that generated it. The actual real-world advancement tables that are available before a tournament are just samples from a process. Sampling error may cause a pick advancement table to violate the Markov assumptions even if the table was sampled from a Markov process. Table 5.2 shows an advancement table that cannot be reverse engineered into a Markov model. However, the mRchmadness simulation method can still be used on a table that cannot be reverse engineered into a Markov model.

Let's see what happens if we try to reverse engineer Table 5.1 into a Markov model of head-to-head pick probabilities using simultaneous linear equations. A partial solution for the Markov model is shown in Table 5.6. The notation P(A beats B) is short for "probability that A is picked to beat B."

The four head-to-head pick probabilities for the first-round matchups in Table 5.6 can be easily determined from the first-round pick advancement probabilities in Table 5.1. Team A has a probability of 0.87 of being picked to advance to the second round and Team D is Team A's opponent. So, the probability that Team A is picked to beat Team D is 0.87. The pick probabilities for the other first-round matchups are determined in a similar manner.

That leaves eight more unknowns. But four of these unknowns can be determined from the others. For instance, P(D beats A) = 1 − P(A beats D). This leaves four possible head-to-head matchups that can occur in the second round.

We know (from Table 5.1) that Team A's probability of being picked to win the second round is 0.61. Consider all the ways that A can win the second round. One way that A can win is to meet B in the second round and defeat B. For this to happen, three outcomes must occur: (1) A is picked to win the first round, (2) B is picked to win the first round, and (3) A is picked to beat B in the second round. Under the Markov assumption, these are independent events. So, we can just multiply their probabilities together. 0.87 × 0.69 × P(A beats B) equals the probability that A meets B and defeats B in the second round.

The only other way that A can win is to meet C in the second round and beat C. 0.87 × 0.31 × P(A beats C) equals the probability that A and C are picked to win their first-round games and A is picked to defeat C in the second round.

TABLE 5.6 Head-to-Head Pick Probabilities Based on Table 5.1

	Team A	Team B	Team C	Team D
Team A		P(A beats B)	P(A beats C)	.87
Team B	1 − P(A beats B)		.69	P(B beats D)
Team C	1 − P(A beats C)	.31		P(C beats D)
Team D	.13	1 − P(B beats C)	1 − P(C beats D)	

These are the only outcomes that have A picked to win in the second round and they are mutually exclusive outcomes, since either B or C must be picked to win the first round. So, we can add the probabilities:

$0.6003 \times$ P(A beats B) $+ 0.2697 \times$ P(A beats C) = the probability that A wins the second round

The probability that A is picked to win the second round is 0.61 according to Table 5.1, so we have:

$0.6003 \times$ P(A beats B) $+ 0.2697 \times$ P(A beats C) $= 0.61$

At this point we have one equation in two unknowns. We can figure out the three other equations using similar reasoning:

$0.6003 \times$ P(A beats B) $+ 0.2697 \times$ P(A beats C) $= 0.61$

$0.6003 \times$ P(B beats A) $+ 0.0897 \times$ P(B beats D) $= 0.28$

$0.2697 \times$ P(C beats A) $+ 0.0403 \times$ P(C beats D) $= 0.06$

$0.0897 \times$ P(D beats B) $+ 0.0403 \times$ P(D beats C) $= 0.05$

It appears we have four equations in eight unknowns. But P(B beats A) $= 1 -$ P(A beats B) , so we can eliminate the unknown P(B beats A) by replacing it with $1 -$ P(A beats B). We can similarly replace three of the other unknowns: P(C beats B), P(C beats A), and P(D beats C). After these replacements, we have four simultaneous equations in four unknowns.

$0.6003 \times$ P(A beats B) $+ 0.2697 \times$ P(A beats C) $= 0.61$

$0.6003 \times (1 -$ P(A beats B)) $+ 0.0897 \times$ P(B beats D) $= 0.28$

$0.2697 \times (1 -$ P(A beats C)) $+ 0.0403 \times$ P(C beats D) $= 0.06$

$0.0897 \times (1 -$ P(B beats D)) $+ 0.0403 \times (1 -$ P(C beats D)) $= 0.05$

TABLE 5.7   Reverse Engineered Markov Model
for Table 5.1 Assuming Maximum Entropy

	Team A	Team B	Team C	Team D
Team A		.637	.844	.870
Team B	.363		.690	.692
Team C	.156	.310		.444
Team D	.130	.308	.556	

So, now we have four equations in four unknowns. We could, in principle, solve these equations by plotting them, but this would require a plot in four-dimensional space, which is hard to visualize. There are several algebraic algorithms for solving simultaneous linear equations. The most general algorithm is Gaussian elimination. But, in the style of Julia Child, we are going to skip some details and jump to the solution.

If we solve these equations, we don't get just one solution. The possible solutions do not constrain the value of P(C beats D) at all. There are valid solutions where P(C beats D) ranges from 0.0 to 1.0. Some additional constraint is needed. One possible constraint is to maximize the entropy of the solution.* The entropy is the sum of the four values $-P \times \log(P)$ when each of the four unknown probabilities is substituted for P. Table 5.7 shows the solution that maximizes the entropy.

The maximum entropy solution can be the best estimator assuming that we have no additional information about the Markov model. It can be better to use additional information if any is available. For instance, the tournament advancement table shown in Table 2.2 was derived from the Markov model in Table 2.1. Table 2.1 is an example of a tournament outcome Markov model from Niemi et al. (2008). Table 2.1 can be accurately reverse engineered from Table 2.2 on the assumption that the probabilities in Table 2.2 are derived based on normal distributions with means equal to the rating spreads (where the spreads

---

* Maximizing entropy was suggested by Mason Wright.

are transitive with respect to addition) and the standard deviation is 11. That is, the assumption is that Table 2.1 was created using Carlin's method since it was cited in the paper and Carlin was an author on the paper. Just maximizing entropy does not accurately reproduce Table 2.1. This additional "information" was really no more than a hunch, but it does illustrate the point that using additional information can improve the accuracy of the estimated Markov model relative to just maximizing entropy.

This demonstrates the method for a four-team tournament. It requires one set of four equations in four unknowns. Applying the method to the full 64-team pick advancement table requires 31 sets of simultaneous linear equations, including 16 sets of four equations in four unknowns to calculate the head-to-head pick probabilities for the second round and one set of 64 equations in 64 unknowns to calculate the head-to-head pick probabilities for the sixth and final round.

Clair and Letscher (2007) derived an ad hoc heuristic method for deriving a Markov model from the pick advancement table for the large 2005 ESPN bracket contest that was available before the tournament. This is the only published method for reverse engineering a Markov model from a pick advancement table. This heuristic does not exactly reproduce the original pick advancement table that was used to calculate it. A simulation based on this Markov model does not have perfect fidelity with respect to the pick advancement probabilities used to calculate it. The simulation is somewhat biased with respect to the original pick advancement table. Clair and Letscher did not use this Markov model in simulations. They only used it in a limited fashion to calculate a subset of the parameters needed to estimate the variance of opponent scores. Most of the parameters in their variance calculation came directly from the original pick advancement table, not from the heuristic method. Their variance calculation itself appears to be relatively robust.

5.7 SUMMARY

Prior to the tournament, limited information is available on opponent behavior in bracket pools in the form of round-by-round pick frequencies for each team, also known as the pick advancement table. Pick advancement tables are not necessarily consistent with a Markov model, and the Markov assumption is not particularly compelling for pick probability models. The mRchmadness algorithm supports simulation of a non-Markov model, but this model requires a different set of simplifying assumptions. Markov models derived from pick advancement tables have been used for bracket improvement (Clair and Letscher 2005, 2006) and have resulted in significant improvement of return on investment (Adams 2017).

The current state of research in this area may not tell the whole story. The best players in your pool who are more likely to compete with you for a prize may not have the same pick patterns as the typical player. But it is unclear if there is any way to exploit such correlations to provide additional bracket improvements. It is even unclear if such possible non-Markov pick correlations are important.

CHAPTER 6

Parametric Whole-Bracket Optimization

T HE MAJORITY OF BRACKETS are submitted to standard scoring pools without upset incentives. TeamRankings.com reports that about 81% of their bracket-advice customer's pools use only round-based standard scoring rules (Barzilai 2015). Clair and Letscher (2007) wrote a research paper that presents a method for optimizing the whole bracket for this most common form of bracket pool. Metrick's method only optimized the champ pick based on simplifying assumptions about earlier round picks.

The Clair and Letscher method is in the class of methods called *parametric methods*. Parametric methods assume that the random variables in the analysis are from probability distributions that are defined by a fixed set of parameters. One of the Clair and Letscher method's basic assumptions is that the score of a bracket is normally distributed. A normal distribution is defined by two parameters: a mean and a variance.

6.1 INPUTS TO THE STRATEGY

Clair and Letscher's paper was the first research paper to present a bracket pool model that includes an opponent model for the current year that is available while it was still possible to submit brackets. Their bracket pool improvement method is based on three inputs: a tournament outcome model, an opponent model, and an estimate of the number of opposing entries.

The development of tournament outcome models is covered in Chapter 2. This model consists of estimated win probabilities for all of the 2016 potential head-to-head matchups that might occur in the 64-team single-elimination NCAA tournament. The tournament outcome model is typically based solely or primarily on each team's performance (win–loss record, margins of victory) for regular season and conference tournament games.

The development of an opponent model is discussed in Chapter 5. Clair and Letscher introduced the strategy of using the pick distribution table from the ESPN (ESPN 2018) or Yahoo (Yahoo 2018) bracket contests as the data source for the opponent model.

An estimate of the number of opposing entries is also required. If you have been playing an office pool, then you can typically base your estimate on the number of opposing entries in previous years' pools.

6.2 THE GOAL: MAXIMIZING EXPECTED RETURN

The goal is to find the bracket that maximizes your expected return. They assumed that you submit a single bracket to your pool, so you pay only one entry fee. The definition that they use for the return on the investment of that entry fee is calculated with this equation:

Return = Standardized Total Return = Winnings/Entry Fee

The return is standardized based on the entry fee, which is also called the *betting unit*. Clair and Letscher analyzed winner-take-all pools where the pot is divided evenly among all winners in the case of a tie for the highest score. If there are 20 opposing brackets then there will be 21 entry fees in the pot, including your fee. If you win the whole pot, then your return is 21 betting units. Your net return (not including your fee) is 20 betting units.

Your average or expected return is calculated based on your probability of winning the whole pot or some fraction of the pot. For each possible score value, the method provides a model that can be used to determine (1) a candidate bracket's probability of ending up with any possible score (2) the expected return for that particular score. The expected return of the candidate bracket is the probability-weighted average, for all possible scores, of the score-specific expected returns. This expected return can be used to compare any two candidate brackets to determine the bracket with the highest expected return. Finally, a search strategy is utilized to estimate the bracket with the highest expected return.

Your expected return is an estimate of how much better your bracket is than the average opposing bracket. If your expected return is one, then you are average. If it's three, then you are three times more likely to win than the average opponent.

In a bracket contest with no entry fee, you can just assume an arbitrary entry fee of unity (1.0) and use the expected return calculation as a measure of how much better your bracket is than the average bracket. Optimizing your expected return will optimize your chances of winning the contest prize.

6.3 ASSUMPTIONS OF THE METHOD

The method requires two assumptions: (1) the score of any specific filled-out bracket is normally distributed (2) all opponents are using the same mixed strategy. Remember that a mixed strategy involves some uncertainty. Clair and Letscher assume that each opponent bracket is a random sample from the opponent model. They assume a Markov probability model for the opponent model, but the method could also be used with the mRchmadness non-Markov model that was discussed in Chapter 5.

In reality, your individual opponents are likely to make few, if any, random picks. In that case, each opponent is using a pure strategy, not a mixed strategy. But you are uncertain of their individual behavior. All you have is a probability model that describes opponent behavior in general. Due to this uncertainty, Clair and Letscher model opponent behavior as if each opponent is using the same mixed strategy that gives results that are consistent with their aggregate behavior.

6.4 THE OPPONENT SCORE PROBABILITY DISTRIBUTION

The key consequence of the two assumptions is that each opponent's score is a random sample from the same normal distribution. This normal distribution has a mean and variance that can be calculated from a tournament outcome model and the opponent model.

Opponent scores are not independent. Opponents have some tendency to pick the same teams. This is reflected in the opponent pick distribution. So, opponent scores will covary together to the extent that they pick the same teams. For instance, almost everyone picks the 1 seed to win in the first round, so scores tend to go up together if the 1 seed wins; this is an example of covarying scores or positive covariance. Opponent score mean, variance, and covariance can be calculated from the tournament outcome model and the opponent model using methods developed by Clair and Letscher that are generalizations of methods developed by Kaplan and Garstka (2001).

6.5 YOUR BRACKET IS THE DECISION VARIABLE

The bracket that you decide to submit to the pool is not modeled as random. These picks are decisions you make. You have to decide on a single bracket that is selected from the approximately 9.2 quintillion candidate brackets. But the score of each candidate bracket can be modeled as a normally distributed random variable. Each has a mean score and a score variance that can be calculated from the tournament outcome model.

Each candidate bracket score covaries to some extent with the typical opponent bracket score. If your bracket has a lot of picks in common with the typical opponent bracket, then the covariance will be high. Also, the magnitude of the score awarded for a correct pick matters. Early round picks don't have a big impact on the covariance since their contribution to the bracket's score is relatively small.

The covariance of a candidate bracket score can be positive or negative. If the bracket goes against the grain and makes picks that are not common opponent picks, then the covariance will be negative. That is, the score of the bracket will tend to move in the opposite direction from the scores of opponent brackets. And, again, the magnitudes of the scores of individual picks matter.

Different candidate brackets have different score means and variances. Each candidate bracket also has a distinct covariance versus the typical opponent bracket. This covariance of a candidate bracket relative to the opposing brackets can be calculated from the tournament outcome model and the opponent model.

6.6 DISTINGUISHING YOUR BRACKET FROM THE COMPETITION

The idea that you need to distinguish your bracket from the competition seems to be common sense, since it's a common piece of advice that you get from many sources. It's the correct strategy. Trying to make your bracket as close to the competition as possible is a breakeven strategy at best.

But the correct approach to distinguishing your bracket is not common sense. The common advice is to bet a lot of upsets. But you can distinguish your bracket by betting no upsets. If you bet the all-favorites bracket, then your expected score will be as high as possible or almost as high as possible. (The expected-point-maximizing bracket will, in rare cases, not pick the favorite over the underdog because the favorite faces relatively stronger competition in earlier rounds.) Clair and Letscher reported that the expected score for the all-favorites bracket was 15 to 20 points above the expected score of opposing brackets for 2005 and 2006.

The all-favorites bracket can be best for a small pool, but for larger pools you need a larger point margin versus the typical opponent in order to win. If there are many opponents, it becomes unlikely to best all their scores with the score of the all-favorites bracket. You have to resort to a risky strategy that has some of the characteristics of the strategy of deliberately fouling when you are behind late in a basketball game. You have to risk a lower score margin in order to increase the probability of getting a score that is high enough to beat all opposing brackets. In statistical terms, you have to increase the variance of your score versus the competition. You do this by betting some upsets.

Picking upsets will change the variance of your score margin against the competition while decreasing your mean score. But you don't want to bet just any upsets. You want to find the upsets that provide the most benefit with the least cost. The benefit is an increase in your score margin variance versus the competition. The cost is a decrease in your expected score.

The touchstone for the quality of advice on betting upsets in a standard scoring pool is the use of an opponent model. If the advice is not based on an opponent model or other quantitative data on opponent picks, then it is not the best advice.

> The best advice for standard scoring pools is based on quantitative estimates of your opponent picks.

6.7 ESTIMATING THE RETURN OF A CANDIDATE BRACKET

To illustrate the process of estimating the return of a candidate bracket, we are going to use the all-favorites bracket as the candidate bracket for a small pool with 20 opposing entries. Except as otherwise noted, we will use a tournament outcome model based on the 2005 Sagarin ratings and an opponent model based on the 2005 Yahoo pick distribution. We will use the exponential scoring rule (1, 2, 4, 8, 16, 32) in specific quantified examples. The data is from Clair and Letscher (2005).

As mentioned earlier, bracket scores covary and therefore are not independent. But, based on simplifying assumptions about the covariance of brackets, Clair and Letscher were able to convert to a probability model where bracket scores are modeled by independent normal distributions. In this converted probability model, each opposing bracket's score is modeled as an independent sample from the same normal distribution. The candidate bracket's score is modeled as a sample from a different independent normal distribution. The mean and variance of these two normal distributions are calculated from the tournament outcome model and the opponent model.

For any given candidate bracket, it is possible to estimate the probability of each possible score. Figure 6.1 shows the probabilities for all possible scores for the all-favorites bracket from the 2005 tournament according to the converted probability model. Figure 6.2 shows the all-favorites bracket. The four games where the lower seed was picked are games where the Sagarin ranking disagrees with the tournament seeding.

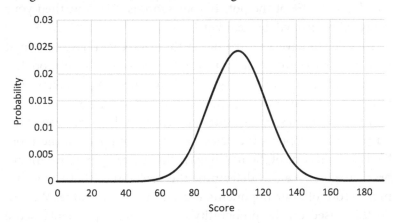

FIGURE 6.1 Probability distribution function of the all-favorites bracket score (based on the independent component of the score variance).

FIGURE 6.2 The 2005 all-favorites bracket. (Reprinted from Clair and Letscher (2005) by permission of the authors.)

It is also possible to represent the expected return for each possible score. If your score is very high, then you are very likely to win the whole pot. Your expected return for a very high score will equal near 100% of the pot. If your score is very low, then your expected return will be near zero.

In order to have a return larger than zero, you have to win at least part of the pot. In order to win the whole pot, your score has to exceed the maximum score of all the opposing brackets. You have to tie this maximum score to win part of the pot.

The maximum score of your opponents tends to increase as the number of opponents increases. The more opponents, the more likely it is that one of them will get lucky and end up with a higher bracket score. We can determine the probability that the maximum score of your opponents exceeds a specific value. We can do this based on the probability that a single opponent's score exceeds that specific value.

The probability that a single opponent's score exceeds a specific value can be determined from the cumulative probability distribution of an opponent score. The cumulative probability distribution of an opponent score is shown in Figure 6.3 for the converted probability model. The graph shows the probability of an opponent scoring below a given horizontal axis score value. For instance, the probability of getting a score below 80 is about 0.4.

Since we are working with transformed probability distributions where the correlations and score dependencies between opponents have been eliminated, we can treat the opponent scores as if they are independent random variables. The event that one opponent's score is below a specific value is independent of the event that another opponent's score is below that value. So, we can apply the multiplication rule for probabilities of independent events. We can multiply the two probabilities together to determine the probability that both scores are below the specific value. Also, the probabilities will be the same for two different opponents since we assumed that all opponent scores are samples from the same normal distribution. So, we calculate the square (p^2) of the probability (p) that one score is below a specific score value to get the probability that both scores are below that score value.

The multiplication rule holds for a large number of opponents. If the probability of one score exceeding a score value is p and

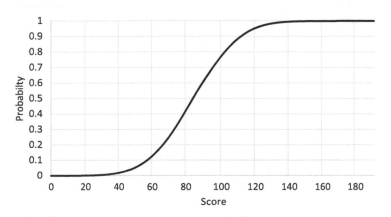

FIGURE 6.3 Cumulative distribution function of an opponent's score (based on the independent component of the score variance).

the number of opponents is equal to N, then the probability that all scores are below the specific value equals p to the Nth power, or p^N. The cumulative distribution function of the maximum score of 20 opponents in the 2005 pool is shown in Figure 6.4. The graph shows the probability that all 20 opponent scores are below a given horizontal axis score. The random variable that represents the highest score is called an *order statistic* because it arises from ordering the scores. The cumulative probability distribution of the highest score (Figure 6.4) is a sigmoid curve, but this is not a normal distribution. Many different types of cumulative distributions have S-shaped curves.

So, for each possible score value of our bracket, we can figure the probability that all opponents score less than our score. This is the probability that we win the whole pot if we have that specific score value.

Also, for each possible score value of our bracket, we can determine the probability that one or more opponent brackets tie our score. We can use the cumulative distribution function to determine the probability that a single opponent's score has a specific value V. The probability that it has the value V is the probability

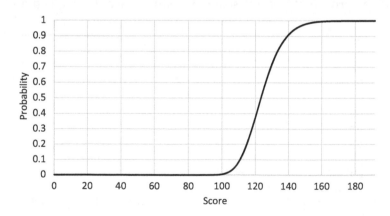

FIGURE 6.4 The cumulative distribution function of the maximum score of 20 opponents (based on the independent component of the score variance).

that it is lower than V + 1 minus the probability that it is lower than V. Using the multiplication rule, we can also calculate the probability that 1, 2, 3, …, N opponents tie our score while all the other opponents have a lower score. So, we can calculate the probability that we win any specific fraction of the pot. Therefore, for any specific score value, we can calculate our expected return if our bracket has that score. The expected return for a range of score values for a 2005 pool with 20 opponents is shown in Figure 6.5.

For any score value, we know the probability that the all-favorites bracket will end up with that score. This probability is represented in Figure 6.1. And, we also know the expected return in a pool with 20 opponents for each score value. This expected return is represented in Figure 6.5. We can now calculate the expected return of the all-favorites bracket in a pool with 20 opponents. The expected return of the bracket is the weighted sum of the expected returns for each possible score. Each score-specific expected return is weighted by the probability that the all-favorites bracket will end up with that score. Figure 6.6 shows a curve representing the values for these probability-weighted expected returns for a range of score values. The area under this curve is the probability-weighted

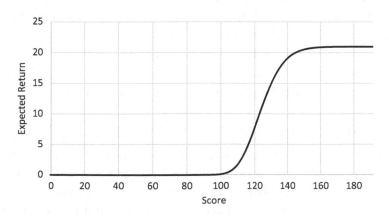

FIGURE 6.5 A bracket's expected return for each bracket score value in a pool with 20 opponents.

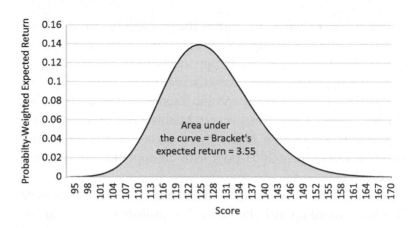

FIGURE 6.6   The expected return of the all-favorites bracket.

sum of the score-specific expected returns. This weighted sum is the expected return of the all-favorites bracket in a pool with 20 opponents. The all-favorites bracket performs well in this small pool, the return is 3.55 betting units. In a pool with a $10 entry fee, your average return would be $35.50. You'd most likely achieve that average by winning the whole $210 pot some years and losing in most years. Assuming ties are rare, you would win about 17% of the time, or less than once in every five years.

We can perform this calculation to estimate the expected return for any candidate bracket. We can compare the expected returns of any two brackets and determine which one has the highest expected return. The bracket with the highest expected return is the better bracket to choose to submit to a bracket pool.

6.8 SEARCHING FOR THE OPTIMAL BRACKET

We now have a way to compare any two candidate brackets and determine which is the better bracket to submit to a bracket pool. We could now calculate and sort the over 9.2 quintillion candidate bracket returns for a particular pool size and pick the best one, but that would be too computation intensive. We need a more efficient search algorithm.

Clair and Letscher used a hill-climbing algorithm. The algorithm climbs via neighboring brackets. Two brackets are neighbors if they are identical except for the outcome of a single game. Since there are 63 games in the tournament, any given bracket has 63 neighbors.

First, the expected return is calculated from some starting bracket and all its 63 neighbors. The bracket with the highest expected return is selected. If the bracket with the highest return is different from the starting bracket, then the process is repeated with the new bracket. The process continues until a bracket is located that has a higher return than all its 63 neighbors. We have reached the top of the hill. But is that just a local maximum? Clair and Letscher experimented with some random starting points and found that the algorithm did not always find the same local maximum, but consistently converged to the same few local maxima.

6.9 THE OPTIMAL BRACKET FOR A POOL WITH A MILLION OPPONENTS

The optimal bracket for a pool with one million opponents is shown in Figure 6.7. In the figure, all the upset picks relative to the all-favorites bracket (Figure 6.2) are circled.

No underdog is advanced to the Sweet Sixteen or to any round before the Sweet Sixteen. The first two rounds are pure chalk. Only one underdog is advanced to the Elite Eight®. Three underdogs are advanced to the Final Four. The stronger remaining teams are advanced after the Final Four. Duke, the only favorite to be picked to make the Final Four, is picked to be the champ. Washington, the stronger team in the matchup, is picked to beat Oklahoma State.

A near optimal bracket may require no more than a few late-round upset picks even in the largest standard scoring pools.

FIGURE 6.7   2005 Estimated optimal bracket for a pool with 1 million opponents with upset picks circled. (Bracket reprinted from Clair and Letscher (2005) by permission of the authors.)

The upsets in this bracket are in stark contrast to common practice and much of typical bracket advice. Many media sources advise people to pick numerous early round upsets and people often follow this advice. In 2005, brackets submitted to the Yahoo bracket contest advanced an average of 11 underdog picks in one of the first three rounds of the tournament. The estimated optimal bracket advanced only one underdog pick.

6.10 THE NORMALITY ASSUMPTIONS EVALUATED

Clair and Letscher used Monte Carlo simulations to evaluate the quality of their normality assumptions. They assumed that bracket score distributions are normal. Also, assumptions about the nature of bracket covariance are required to arrive at the converted probability model where the normal distributions for the opponent brackets and candidate bracket are independent.

In their first experiment, they looked at the quality of the assumption that opponent scores are normally distributed. They compare three opponent score distributions. The first distribution is based on a sample of 5000 real brackets (submitted by humans) sampled from the 2004 ESPN Tournament Challenge. The second distribution is based on 5000 brackets simulated using a Markov opponent model derived from brackets submitted to the 2004 ESPN Tournament Challenge. The third distribution is the normal distribution of opponent scores estimated using their parametric method. In all three cases, the same tournament outcome model is used to simulate bracket scores or as input to derive parameters for the parametric method. The bracket scores are estimated based on 10,000 tournament simulations.

The 2004 distribution of the simulated opponents is very close to the distribution of real opponents. Relative to the estimated normal distribution of opponent scores, the distribution of real opponents is slightly skewed with a slightly fatter tail to the right (toward higher scores). This suggests that using a skewed distribution with a fatter right tail might improve the estimation.

Clair and Letscher indicate that the skewness is due to high point totals in later round games. The last few games are relatively large contributors to the bracket score given the typical top-heavy scoring rules. And, there are only a few of these games, so the law of large numbers and the central limit theorem (which are the justifications for using a parametric approximation based on the normal distribution) don't hold too well for the score variations due to the outcome of the last few games.

In their second experiment, they look at the quality of the expected return estimates. They estimate the expected return of 100 real brackets submitted by humans to the 2004 ESPN Tournament Challenge. They calculate expected return for a hypothetical pool with 100 opposing brackets using three different methods. The first method is based on 100,000 tournament simulations where the opponents are samples of real opponents. The second method is based on 100,000 tournament simulations using simulated opponents from a Markov opponent model. The third method is their parametric method.

For the set of 100 candidate brackets, returns from the normal approximation method correlated with an R-squared of 0.53 with the returns estimated from simulations using real opponents. This indicates that the normal approximation captures 53% of the variation in the returns estimated using real opponents. The returns estimated from simulated opponents captured 98% of the variation in the returns estimated using real opponents.

The results of the second experiment indicate that simulations provide very accurate return estimates for candidate brackets. But simulations are limited in their range of application. They were too computation intensive to apply to bracket contests with millions of brackets in 2007 and that is probably still true today. The Clair and Letscher method is the only publicly available method that provides estimates of the optimal brackets for these large bracket contests.

Overall, the results are quite supportive of the use of an opponent model in simulations. But the opponent model used here was derived directly from a large sample of brackets, not from the pick distribution that summarizes the picks for each round. So, the results are not directly supportive of the opponent models derived from the pick distribution.

These experiments don't evaluate the quality of the tournament outcome model. The same tournament outcome model is used in all the methods that are compared.

6.11 VARIATIONS IN TOURNAMENT OUTCOME MODELS

Clair and Letscher calculated results based on three different tournament outcome models for the 2004 and 2005 tournaments. The tournament outcome models were based on the following three different sources: Jeff Sagarin's ratings, Ken Massey's ratings, and historical seed performance statistics from past tournaments. The optimal brackets were similar in the early rounds. The picks for the 2004 Final Four varied a good bit, with all three models agreeing on only one pick. There was more agreement in 2005. In 2005, all picked a Duke versus Washington final.

When the return of the optimal picks of one tournament outcome model was evaluated with a different tournament outcome model, there was a large disagreement. The 2004 optimal Massey bracket had a return of 798.8 for Massey, but only a 363.5 return for Sagarin and a 60.8 return for historical seed performance. (These are estimated expected returns for a pool with 5 million entries.) But the disagreement was more due to a large general disagreement about the return of any candidate bracket in 2004. All models tend to agree that the optimal bracket, according to one tournament outcome model, was reasonably close to optimal according to another tournament outcome model.

Note that these variations due to the selection of the tournament outcome model are not just a characteristic of the normal approximation method. Such variations are likely for any bracket optimization method. Clair and Letscher speculated that this tournament outcome model dependency might be mitigated if the expected return of a set of multiple entries is optimized. They did not extend the method so that it could be applied to multiple entries. The rules of many bracket pools allow more than one entry per participant.

6.12 SOURCES FOR THE OPPONENT MODEL

Clair and Letscher used a large sample of brackets from real opponents to estimate their 2004 opponent models. For 2005 and later years, they used an ad hoc heuristic method to compute opponent models from the tournament pick distribution. The latter method can provide an opponent model that can be used for bracket improvement before betting closes in the typical office bracket pool.

The opponent model that results from this heuristic provides somewhat biased estimates of the pick advancement probabilities versus those of the source tournament pick distribution. Head-to-head pick probabilities from this opponent model were used only in the estimation of the variance of opponent scores. Chapter 5 covered some other methods for calculating opponent models that have exact fidelity to the pick advancement table. These alternative opponent models would provide different estimates of the variance of opponent scores that might be considered theoretically more pristine. But it's hard to say if this would be a practical improvement in the Clair and Letscher method. As pointed out in Chapter 5, there are typically a range of different opponent models that have exact fidelity to the pick distribution, so using a reverse-engineered opponent model in the Clair and Letscher method would not necessarily improve the estimate of the variance of opponent scores.

6.13 CONCLUSION

Clair and Letscher introduced a full-bracket probability model for pools. They developed the only full-bracket parametric method publicly available for estimating the return on investment of a bracket submission to a pool.

Their results generally support the findings and conjectures in Andrew Metrick's earlier paper on standard scoring pools. Metrick's method tended toward somewhat more contrarian champ picks because the estimates were based on the simplifying assumption that a chalky down-bracket was no better than your typical opponent's down-bracket. The Clair and Letscher method is relatively more likely to pick one of the less popular 1 seeds for champ (rather than picking a 2 seed for champ) because it estimates that the optimal down-bracket is superior to the typical opposing down-bracket.

Clair and Letscher's results indicate that, even in bracket contests with a million entries, most early round upset picks are a bad idea, and only a few upsets are required in the later rounds to produce an optimal bracket.

CHAPTER 7

A Practical Contrarian Strategy

J ARAD NIEMI AND HIS co-authors (Niemi et al. 2008) analyzed the brackets submitted to three pools held in the years 2003 to 2005. More details of the research are provided in Niemi's master's thesis (Niemi 2005). The pools were located in the Chicago area. These were standard scoring pools using exponential (1, 2, 4, 8, 16, 32) scoring, the most common bracket pool scoring rules. They estimated the net expected return on investment (ROI) for all these brackets using simulations based on four different tournament outcome models. They used these ROIs to evaluate bracket improvement metrics and strategies.

7.1 POOL BETTING BEHAVIOR

Basic statistics from these three pools are presented in Table 7.1. The 2003 and 2005 brackets evidenced the typical pattern, first reported by Metrick: the 1 seeds are over-backed.

In 2003, 51% of the brackets picked Kentucky as champ. The tournament outcome models gave Kentucky at most a 17% chance of winning. 2004 was a bit of an oddball year. Two seed Connecticut was the most picked champ and 2 seed Oklahoma State came as

TABLE 7.1   Exploratory Data Analysis of Chicago Office Pool Sheets

Year	2003	2004	2005
Participants	113	138	167
Champions Bet by Seed			
1	86 (76%)	61 (44%)	137 (82%)
2	14 (12%)	58 (42%)	18 (11%)
3	4 (4%)	10 (7%)	5 (3%)
4	7 (6%)	3 (2%)	4 (2%)

From Niemi et al. (2008), copyright © American Statistical Association, www.amstat.org, reprinted by permission of Taylor & Francis Ltd, http://www.tandfonline.com on behalf of American Statistical Association.

the third most picked champ. In 2005, 49.7% of the brackets had Illinois as the champ pick, but Illinois had a win probability in the range of 9% to 40% depending on which tournament outcome model is used. Sagarin Predictor gave Illinois a low win probability, whereas Sagarin Elo gave Illinois a high win probability. The excessive pick rate for Illinois represented a bias toward the home state team in this Chicago-based pool. Illinois' pick rate in the nationwide Yahoo bracket contest was 31.5%.

7.2 DEFINING RETURN ON INVESTMENT

The return on investment (ROI) value used by Niemi is net return rather than total return. The formula is:

ROI = Standardized Net Return = (Winnings – Entry Fee)/Entry Fee

The Standardized Net Return is one betting unit less than the Standardized Total Return used in the research presented in the previous chapter. Winnings were calculated based on the payout structure of the pools. The payouts varied from year to year and were roughly 45%, 22.5%, 15%, 10%, and 7.5% of the pot for 1st through 5th places, respectively.

7.3 TOURNAMENT OUTCOME MODELS

Three of the tournament outcome models were based on three Sagarin ratings: Predictor, Elo, and the overall Sagarin Ratings.

The Predictor is based purely on point margins. Elo is based purely on wins and losses with no regard to point margin. The overall Sagarin rating was a composite of Predictor and Elo ratings during this time period. The fourth rating used was the Vegas rating, a rating described by Kaplan and Garstka (2001). This rating is based on betting market point spreads and point totals that are available for the 32 first-round games prior to the tournament. Bookmakers take bets on the total combined points scored by both teams in a basketball game (also called the "over/under") in addition to the point spreads. As with the point spreads, the betting market total constitutes a crowdsourced estimate of the sum of the total points scored by both teams in a game. The Vegas rating of a team equals a team score estimate calculated from the betting market spread and total for the team's first-round game. The Vegas rating of the favorite is estimated as:

Favorite's Rating = (Total + Spread)/2

Underdog's Rating = (Total – Spread)/2

These equations come from solving two simultaneous linear equations that represent the mathematical definition of the spread and the total.

Favorite's Score – Underdog's Score = Spread

Favorite's Score + Underdog's Score = Total

With the ratings from these four rating systems in hand, the four tournament outcome models are derived using the methods described in Chapter 2. For each possible head-to-head matchup that may occur in a tournament, the difference in the ratings of the two teams is assumed to be an estimate of the mean point margin of the game, and the actual point margin is assumed to be a normal distribution (a bell curve) around that mean. The area under the bell curve up to zero represents the underdog's win probability for this particular head-to-head matchup. The probabilities for all possible head-to-head matchups between the 63 teams in the tournament constitute the tournament outcome model for a particular rating system.

7.4 ESTIMATING ROIS USING SIMULATIONS

The ROIs of the brackets in the three pools are estimated using simulated tournaments. In the simulations, the brackets entered into the pools are known quantities; they were the brackets that the real people actually submitted to the pools. Since brackets submitted to the pools are treated as a known quantity, the only unknown is the statistical distribution of tournament outcomes and the resulting statistical distribution of pool winners.

This is a different approach to accounting for opponent behavior. Fifty-eight brackets picked Kentucky for champ in the 2003 pool. The pool has 113 total brackets submitted. Niemi directly uses these 113 brackets to represent opponent behavior. Metrick's approach (as represented in his paper) for this data would be to assume opponents had a probability of 58/113 of picking Kentucky and analyze the pool on that basis. Clair and Letscher used a number of approaches including sampling the brackets of real people and simulating brackets based on a derived probability model.

The tournament is simulated using the tournament outcome model. The outcome of each first-round game is simulated, using a uniform random number generator, based on its outcome probability in the model. A uniform random number generator produces a random number uniformly distributed over values between 0 and 1. Sixty percent of the time, the random number is below 0.6. So, if the win probability of the favorite is 0.6 then we have a simulated win for the favorite when the random number is below 0.6. We can do this for all the first-round games to determine the second-round matchups. We can iterate through all six tournament rounds to determine the outcome of all games. We have simulated the tournament. We can now determine the scores awarded to all brackets. We award the prizes to the winning brackets. We have now fully simulated one pool and determined the ROI of every bracket in the pool for that one tournament simulation.

Then we iterate the process, simulating the tournament all over again and determine the bracket ROIs for that simulation. In the first phase of this research, Niemi simulated 1000 tournament outcomes for each of the four tournament outcome models. For each bracket, averaging the 1000 ROIs calculated from the 1000 tournament simulations based on a single tournament outcome model provides an estimate of the bracket's ROI, assuming that the tournament outcome model is correct. This process provides four estimated ROIs (one for each tournament outcome model) for each of the 418 brackets submitted to the three pools.

7.5 MOST PLAYERS MAKE BAD BETS

The results show that, for all four tournament outcome models, 60% or more of the brackets submitted to these pools have an estimated ROI that is less than zero. These brackets are in need of improvement just to be breakeven bets. Depending on the year and the tournament outcome model used, somewhere between 7% and 16% of the brackets were good enough to have an estimated ROI greater than one. These brackets are estimated to double your money on average. Some brackets had higher estimated ROIs. In 2003, one player had an estimated ROI exceeding 3.5 according to the Predictor-based tournament outcome model.

7.6 A SIMILARITY METRIC

Niemi et al. developed a similarity metric and evaluated the metric's ability to separate profitable contrarian brackets from money-losing brackets. The similarity metric is calculated for each game and then all the game metrics are summed to get the overall bracket's similarity score. For a given game, the game's raw similarity score is the number of points awarded for correctly picking the game multiplied by the proportion of your opponent's brackets that pick that game. So, if you picked the most popular team, Illinois, to be champ of the 2005 pool, the raw similarity score for that game would be 0.497 × 32 = 15.9, since Illinois was picked for champ by 49.7% of brackets and the pool scoring rules award 32 point for correctly picking the champ. Picking North Carolina as champ reduces the raw game similarity score to 0.228 × 32 = 7.3 since North Carolina was picked for champ by 22.8%. The raw game scores are summed to get an overall raw bracket score. Then the raw score is normalized so that the similarity metric's value must fall in the range between 0 and 1. The score is normalized by dividing it by the maximum possible score that a sheet can get. The maximum for the exponential pool scoring rules is 192. Your bracket would have a similarity metric value of 1 only if it was perfectly identical to all the other brackets in the pool and this could happen only if they were all identical to each other. As a practical matter, the maximum similarity metric value is likely to be less than 0.7.

Figure 7.1 is a contour plot based on three characteristics of each of the 418 brackets from the three years of the Chicago pool under analysis. Bracket similarity is plotted on the vertical axis. Bracket ROI is indicated by shading. The darker the shading, the higher the ROI. The lightest two shades of grey indicate brackets that are estimated to lose money; this is the region that you surely want to avoid. The horizontal axis plots the natural log of probability that all the games in the bracket will be correctly picked. This is the probability that the bracket could win Warren Buffet's famous contest for the perfect bracket. You can look at the horizontal axis as a measure of the chalkiness of a bracket. The chalkier brackets that have the higher expected bracket pool score will be to the right. The probabilities are calculated using the Predictor tournament outcome model.

As you can see from the plot, brackets with the highest similarity for any given log probability are all losers. All the pool sheets that have a 0.62 or higher similarity are losers. Over most of the plot, moving down or to the right improves ROI. At least, that is true for brackets with a log probability above −30. So, for a given game, picking a relatively stronger team that is not relatively more popular will improve your bracket. And picking a relatively

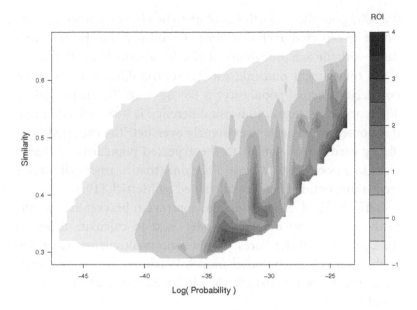

FIGURE 7.1 Contour plot of ROI by similarity and log probability. (From Niemi et al. (2008), © American Statistical Association, www. amstat.org, reprinted by permission of Taylor & Francis Ltd, http:// www.tandfonline.com on behalf of American Statistical Association.)

less popular team that is not relatively weaker will improve your bracket.

7.7 IDENTIFYING CONTRARIAN CHAMPS

Niemi compared the ROIs of expected-point-maximizing (EPM) brackets and contrarian brackets. The EPM bracket, which is discussed in Chapter 4, is the bracket with the highest projected average pool score. The contrarian bracket is the EPM bracket with the champ pick replaced by the team that is the most under-bet for champion. The most under-bet team for champ is determined by the difference between the number of brackets that chose the champ and the number of brackets that would be expected to choose the champ based on the probability that the team wins the championship. For instance, 38 of the 167 brackets submitted to the 2005 pool chose North Carolina to be champ. But 56 brackets would be expected to choose North Carolina if they bet in accordance with the win probability of North Carolina in the Predictor-based tournament outcome model. So, the difference (expected popularity – actual popularity) is 56 – 38 = 18. This is the largest difference of all teams. Illinois's difference is −68 indicating that this hometown favorite is extremely over-bet. The expected popularity used here is similar to the expected popularity of champ picks in a pool at financial equilibrium (that is, where all players are making optimal bets) as calculated by Metrick (1996).

Both the EPM bracket and the contrarian bracket are specific to the tournament outcome model used to calculate them. In the 2005 pool, Wake Forest is the contrarian champ according to the overall Sagarin-based model. Washington is the contrarian champ according to both the Elo-based model and the Vegas-based model.

Niemi used simulations to estimate the ROIs of the contrarian brackets and the EPM brackets for all four tournament outcome models for the pool for each of the three years. In all cases, the ROI of the contrarian bracket exceeded or equaled the ROI of the EPM bracket. In four cases, the contrarian bracket was the same as the EPM bracket. Interestingly, the contrarian bracket was the same as the EPM bracket for the Predictor-based model for all three years. In these cases, you would not need to pick a weaker champ to be the contrarian champ, the champ in the contrarian bracket is also the champ in the EPM bracket. You'd be picking the most likely winner for all of the 63 games in the tournament. All the contrarian bracket ROIs exceeded 2 indicating that you would be expected to triple your money.

The evidence presented in Chapter 4 indicates that the EPM bracket performs very well in pools with upset incentives. But for standard scoring pools where no bonus points are awarded for upsets, the evidence indicates that a contrarian bracket will improve your ROI or, at worst, leave your ROI unchanged.

7.8 IMPROVING YOUR BRACKET

You cannot use Niemi's specific method directly to compare team strength with team popularity because you typically don't have access to your opponent's brackets before you have to make your bets. But you can use the ESPN or Yahoo contest pick distribution tables (ESPN 2018; Yahoo 2018), perhaps adjusted for hometown bias (Null 2016) as a proxy to estimate the tabulation of your opponent's champ picks. Or, you can let Chris Wilson do the work for you (Wilson 2018). Wilson interviewed Jarad Niemi and some other bracket pool experts in 2009 and wrote a web article where he estimated the best contrarian picks for champ. He's been writing essentially the same (updated for the current year) article every year since, using a method similar to Niemi's. He headlines with his best contrarian pick for champ and mentions a number of other good contrarian picks. For over 10 years, he has headlined only one contrarian champ that actually won the tournament (Duke in 2010), but one win in 10 years is enough to achieve a positive ROI in all but the smallest pools, those with less than 10 competing entries.

> Comparing each team's win probability with its pick probability can determine a contrarian champ pick for a bracket with a good ROI.

You will need an EPM down-bracket along with a contrarian champ pick. The simplest way to get something close to an EPM down-bracket is to simply bet the lower (stronger) seed in each round. Or, you can pick the stronger teams according to any good predictive rating system. Or, pick the stronger team according to the betting market spreads for the first round. EPM brackets and tournament advancement tables are provided by FiveThirtyEight.com and Poologic.com. Kenneth Pomeroy provides a tournament advancement table based on his rating (Pomeroy 2018).

7.9 CONCLUSION

Niemi et al. show that the best brackets evidence relatively low similarity with competing brackets while also maintaining relatively high overall chalkiness as indicated by their probability of correctly picking all the games. This research also provides evidence that a practical procedure, supported by readily available data sources, can be used to create a contrarian bracket that has a positive ROI and that performs as good as or better than the EPM bracket in standard scoring pools.

CHAPTER 8

Statistical Hypothesis Testing

L ET'S SAY YOU HAVE been entering one bracket in a winner-take-all pool with 30 entries and a $10 entry fee for the last 10 years. You have won in one of those years. That puts you well into the black since you have won $300 and all your fees add up to $100. But, does your success actually constitute convincing evidence that you have a superior bracket strategy?

If you were an average player, you would have a 1/30 chance of winning. (We are assuming that the pool has a rule for breaking ties, so we can ignore ties.). Your chance of losing is 29/30. The outcomes from year to year are presumably independent events so you can multiply their probabilities together. Your chance of losing in all 10 years is $(29/30)^{10}$, which comes out to about 0.71 or 71%. Your chance of winning one or more pools is about 29%, a little less than one chance in three, even if you have no superior skill. Winning only once is not that impressive. This or better would happen by chance about one-third of the time even if you are just an average player.

This reasoning constitutes a *statistical hypothesis test*. In the terminology of statistical hypothesis testing, we have failed to reject the hypothesis that you are an average player in this pool. This hypothesis (the hypothesis that you would like to reject) is called the *null hypothesis*.

You might want to claim that this test proves nothing, that you are indeed a superior player. The hypothesis that you are a superior player is called the *alternative hypothesis*. If you are right, then we failed to reject a false null hypothesis. This is called a *type II error* or a false negative. (The *type I error* or false positive is when we reject a true null hypothesis.)

8.1 STATISTICAL POWER

Perhaps you could silence the scoffers via another hypothesis test. You could do another 10-year test. If you played one entry for the next 10 years maybe you would win more than once. If you won more than once in the next 10 years, would that allow us to reject the null hypothesis? The math is a bit more complicated for this case, but the calculations can be performed using the *binomial test*. The binomial test is appropriate because the average player's wins can be modeled as a binomial probability distribution. The binomial distribution represents the probability of any number of successes for a fixed number of trials and a fixed probability of success. In our case, we have 10 trials (pools) that result in either a win or a loss and the win probability of the average player is 1/30.

The binomial test shows that an average player in a pool with 30 entries would have a 0.042 probability of winning two or more times in 10 years. That would be pretty convincing evidence that you are a superior pool player by the usual standards of hypothesis testing. Anything below a *significance level* of 5% is typically considered to be a good basis for rejecting the null hypothesis. But it also depends on how credible the alternative hypothesis is in the first place. If you are picking your teams based on which mascot is the strongest, or if you claim to have extrasensory perception, then people might still think you were just lucky. But, if you were using bracket improvement strategies based on the peer-reviewed research, then the notion that you are a superior pool player would already be plausible.

But what are the chances that you will win twice in the next 10 years? We don't have enough information to calculate that. How much better are you than the average player? We have to know the answer to that question, or at least have an estimate. If you are using the strategies from the peer-reviewed literature, then it's reasonable to estimate that your win probability is around three times better than that of the average player. This quantity (three times better or 300%) is a measure of the *effect size* of a strategy. The inverse (1/3) is the *relative risk* of losing and is also called the *risk ratio*. A good strategy is protective against the risk of losing. Relative risk is one of many different metrics that may be used to quantify effect sizes. A particular effect size metric is chosen based on its usefulness for communicating the magnitude of the effect or improving the experimental design. In our example, relative risk is useful for both purposes.

The average player wins 1 in 30 pools on average. If you are three times better, then you will win 1 in 10 pools or 10% of the time on average. Using the binomial test calculation, your chances of winning two or more pools over 10 years is about 0.26 or 26% of the time. The 0.26 value is called the *power* of the statistical hypothesis test. The power of a statistical test is the probability that the test will result in the rejection of the null hypothesis.

So, you will wait 10 years for a new hypothesis test to complete, and there is only a 26% chance that it will reject the notion that you are just average. And this, even assuming that you are three times better than average!

The problem is that the proposed statistical test is weak. How can we make it stronger? One thing to try is to enter more brackets. Let's see what would happen if you entered 15 brackets in a pool with 30 opposing brackets. Of course, your pool might have a rule that limits the number of brackets that you can enter. Or the pool czar might balk when you try to do this even if there is no explicit rule against it. But you could always do a "what if" experiment where you prepare 15 brackets before betting closes and see how they would have done if you had been able to submit them.

The null hypothesis will now be that your 15 brackets are overall average. There will be 30 opposing entries and 15 of your entries for a total of 45 entries. You will be betting 1/3 of the entries. If the null hypothesis is true, then you have a 1/3 chance of winning the pool. According to the binomial test calculation, the chance that you will win in six or more of the 10 years is 7.6%, so that would not be good enough to reject the null hypothesis at the 5% significance level. You'd have to win seven or more times in 10 years. The probability of that happening under the null hypothesis is less than 2%.

What is the power of this test? If you win three times better than average then you will win all 10 years, but that is not a realistic assumption. You have been submitting the best single bracket that you could estimate, so those other 14 brackets have to be less likely to win. We need a new estimate for the effect size of betting 15 brackets instead of just one bracket. Let's assume that 200% or two times better than average is a reasonable estimate of the effect size. Fifteen average brackets have a 1/3 chance of winning, so your superior set of 15 brackets will have a 2/3 chance of winning. If you have a 2/3 chance of winning, then your chance of winning the required seven or more pools is about 56% according to the binomial test calculation. The power of the test is 0.56.

So, a statistical hypothesis test where you submit one bracket has a power of 26%. But if you submit 15 brackets, the power goes up to 56%. The 56% is not a slam dunk, but it's lot better than 26%. Using some basic statistical tools and a few assumptions, we have come up with a better design for a statistical hypothesis test.

The rest of this chapter will describe an actual statistical hypothesis test using 10 years of pool data from a standard scoring pool. This pool has four payouts. The binomial test is not applicable to this data because the binomial test is only applicable to binary outcomes. The outcome of a pool with four payouts is not just a win–lose proposition. The returns for someone playing 1/3 of the total entries for 10 years in a pool with four payouts are roughly normally distributed, so we can invoke the central limit theorem and use a statistical test that assumes normal distributions. But first we need to come up with an effective multiple-entry strategy.

8.2 BACK-TESTING A MULTIPLE-ENTRY STRATEGY

Strategies for submitting multiple entries had been little discussed in the peer-reviewed literature. Breiter and Carlin (1997) referred to the "time-honored method of multiple pool entries." They suggested the heuristic of using multiple entries from different strategies for optimizing brackets for submission to pools with upset incentives. Clair and Letscher (2007) pointed out that submitting multiple entries could reduce dependence on a specific tournament outcome model. And Breiter and Carlin pointed out that the variability in the tournament outcome and the bracket scores is large even if the tournament outcome model specifies each probability correctly. The precision of the estimated score of a single optimized bracket is low even if we assume that the accuracy is high.

In 2017, I presented research on multiple entry strategies in standard scoring pools at the poster session of the New England Symposium on Statistics in Sports (Adams 2017). The goal of this research was to test the quality of return on investment (ROI) estimates from a bracket strategy using a statistical hypothesis test.

A multiple entry strategy was back-tested on 10 years of data from an office pool located in Connecticut. The back-test involved predicting the best brackets to submit to these pools and then determining how these brackets would have performed if they had been submitted to these historical pools.

Historical pool data was available for the years 2008 through 2017. This was a standard scoring pool using exponential scoring (1, 2, 4, 8, 16, 32). The number of brackets submitted ranged from 178 to 241. The entry fee was five dollars. The pool awarded four prizes: 40%, 30%, 20%, and 10% of the pot. In the actual pools, ties were broken based on the best guess of the total score in the championship game, a guess had to be submitted with each bracket entry. But in this ROI analysis, the winnings were assumed to be divided in the case of a tie. The betting pattern in these pools had the characteristics noted in earlier research. The 1 seeds are over-bet. The hometown effect is evident: there is a home state bias in favor of the Connecticut Huskies.

8.2.1 Calculating Bracket ROIs

Bracket ROIs were estimated using pool simulations. These used the probability model of a bracket pool described by Clair and Letscher (2007). The inputs to modeling a pool are (1) N, the number of opponent entries, (2) a tournament outcome model, and (3) an opponent model. The tournament outcome is a random variable based on the tournament outcome model. Opponent brackets are random variables based on the opponent model. One pool simulation consists of one tournament outcome and N opponent brackets. The tournament outcome model was based on the Sagarin Predictor ratings. The opponent model was derived from either the Yahoo pick distribution (Yahoo 2018) or ESPN pick distribution (ESPN 2018) using the heuristic from Clair and Letscher's research paper to estimate the head-to-head pick probabilities for each game. No adjustments to the pick distributions were made to address the home state bias.

As mentioned in Chapters 6 and 7, this heuristic for deriving an opponent model is somewhat biased. Later review of the results indicated some bias: simulated opponents were somewhat less likely to advance the most popular champ pick than they should have been according to the pick advancement table for the nationwide bracket contest. A reanalysis performed using the mRchmadness method for deriving the opponent model indicated a slightly higher mean and a slightly higher variance of the ROIs. But overall the results were not greatly improved by changing the opponent model.

The number of opponent entries used in these simulations was 200. In actual practice, the specific number of pool entries is not known in advance, so the number has to be estimated. Since pre-2008 pools had approximately 200 entries, 200 is a reasonable estimate. Bracket optimization results are relatively insensitive to errors in the estimated number of pool entries.

All the inputs used in this bracket pool probability model are available before tournament tip-off during the period while it is still possible to submit brackets to a pool. The Sagarin ratings for the season are available on Selection Sunday. The Yahoo and ESPN pick distributions are available and updated multiple times before the Thursday tournament tip-off. The number of opponent entries is estimated from past pools.

The bracket ROIs were based on 10,000 pool simulations for each year. For each simulated pool, the tournament outcome and all opponent scores that were among the top four scores were saved. This is the minimum information needed to determine the ROI of any additional brackets that may be added to the pool. This information allows us to estimate the ROI of any candidate bracket that we are considering for submission to the pool.

The ROI of a candidate bracket in a simulated pool is calculated by computationally "submitting" it to the pool and determining how it would have fared in the pool competition. The bracket is scored based on the tournament outcome for that pool. If the brackets score is not among the top four scores in the pool, then its ROI is minus five dollars (minus one betting unit). That is, you just lost your entry fee. If the bracket score is above the highest opponent score, then the bracket ROI will be 40% of the pot minus the entry fee, or $201 \times 0.4 - 1 = 79.4$ betting units or $397. If the score falls among the top four opponent scores, then it's ROI depends on the bracket score rank and whether it ties one or more scores. The estimated average ROI of a bracket is the bracket's average ROI over the 10,000 simulated pools.

The estimated average ROI or expected ROI of a set of candidate brackets submitted to the pool can be calculated in a similar manner. If a set of 10 brackets are submitted and none are in the money, then the ROI is −10 betting units. If one has the highest score and none of the others are in the top four scores, then the ROI is $210 \times .4 - 10 = 74$ betting units. The estimated expected ROI of the set is the average ROI for the 10,000 simulated pools.

8.2.2 Optimizing ROI

Clair and Letscher optimized the ROI of a single candidate bracket using a hill-climbing algorithm. It is necessary to generalize this process to mutually optimize multiple brackets.

For the multiple-entry strategy, an iterative algorithm is used. First, a single bracket is optimized using a hill-climbing algorithm. A commitment is made to submit this optimized bracket to the pool. Then the next bracket is optimized using the hill-climbing algorithm. But this next bracket is optimized relative to the pool consisting of the opponent brackets plus our previously submitted optimized bracket(s). So, each additional bracket is optimized using knowledge of our previous submittals.

The hill-climbing algorithm was used to optimize picks for only the last seven games of the tournament. These seven games are the four regional championship games, the two semi-final games, and the final championship game. The optimization was limited to the last seven games of the tournament in order to reduce the computation time for what is already a computation intensive algorithm.

For each candidate bracket, the hill climb always starts with the same bracket. This is the bracket where the higher seed is picked to win each game. Then, starting with the championship game, different teams are tried as picks for that game and the bracket's expected ROI is calculated for each team. The bracket ROI is maximized over a range of possible picks for that game. Some of the lower-seeded teams are not considered to reduce computation time since these lower-seeded teams are very unlikely to advance deep into the tournament. After the champ pick is optimized, the runner-up is optimized in the same manner. Then the two other regional champs are optimized. Finally, the process is repeated again, starting with the champ. The process is repeated until repetition provides no improvement in the ROI.

This hill-climbing algorithm is different from the one that Clair and Letscher employed (which is described in Chapter 6). The Clair and Letscher algorithm was not tested in this application, so it is unclear if their algorithm would be more accurate or less computation intensive in this application.

8.2.3 Results

Figure 8.1 shows results of entering 100 mutually optimized entries in the 10 bracket pools. The ROIs ranged from 209% to −100%. The whole pot was won in the best year and the pool had over 200 entries, resulting in an ROI of 209%. In the worst year, there were no winnings leading to a 100% loss of all entry fees.

The two years when the multiple entry strategy failed to show a profit were years when the Connecticut Huskies won the tournament. Since the office was located in Connecticut, there was

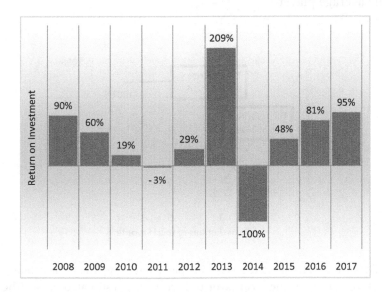

FIGURE 8.1   The ROI of 100 mutually optimized pool entries in a pool with approximately 200 opponent entries.

a large home state bias (see Figure 8.2) in the pick distribution. The home state bias is similar to estimates by Null (2016). Even a multiple entry strategy that involved betting approximately 1/3 of the pool entries could not overcome this bias when the home state team won the tournament. Adjusting the opponent model for home state bias using Null's estimates is probably a good idea in general, but it would not have prevented these losses. Such an adjustment would not help the strategy in a year when the home state team wins because the more over-bet teams tend to be advanced less in the optimized brackets. Multiple entry strategy for a single pool does not mitigate the risk of the home state team winning the tournament.

Over the 10 years of the back-test, the multiple entry strategy had an average ROI of 53%. The standard deviation of returns was 79%. The p-value of a one-tailed T-test is 3.2% indicating grounds for accepting the hypothesis that this large-scale, multiple-entry strategy is superior to the strategy employed by the average player.

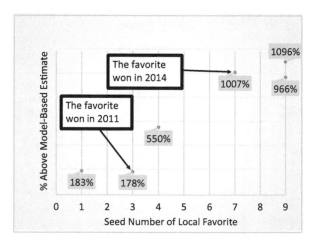

FIGURE 8.2 Home state bias: The percentage by which the champ pick frequency of the local favorite exceeded its estimated value. The estimate was based on the pick advancement table from a nationwide bracket contest.

8.3 PRACTICAL MULTIPLE- ENTRY STRATEGY

Your pool czar will probably not let you bet one-third of the brackets in your pool. You can generate a few brackets for multiple entries by varying your champ pick. Pick from among the strongest teams according to a predictive rating system. Chapter 10 provides information on websites that provide advice on multiple entry strategies informed by bracket pool simulations.

> Varying your champ pick is a good way to generate multiple entries for a bracket pool.

CHAPTER 9

Psychobracketology

Rational choice theory was dominant in economics in the 1950s. But people did not always seem to act in accordance with their rational economic interest. Metrick showed this in his analysis of the bracket pool. The Nash equilibrium calculation, used by Metrick, provided a way to measure the deviation of economic behavior from the rational model. This deviation was sometimes dubbed a "non-pecuniary payoff." That was the only interpretation that was consistent with rational choice theory. Economic man was rationally seeking an optimal payoff that was not always measured in dollars and cents. This often seemed the correct interpretation and it could be difficult to refute.

An arguably more effective attack on the centrality of rational choice theory came from another source. It came from certain psychological experiments. Even back in the 1950s, the cracks were evident. Experiments in psychology laboratories on a phenomenon called probability matching were getting the attention of economists, and the implications seemed to be at odds with rational choice theory.

9.1 WHAT IS PROBABILITY MATCHING?

Tests similar to the following one have been used in research on probability matching:

A die with four red faces and two green faces will be rolled 90 times. Before each roll you will be asked to predict which color (red or green) will show up once the die is rolled. You will be given one dollar for each correct prediction. Assume that you want to make as much money as possible. Consider these two strategies:

Strategy A: Make predictions according to the frequency of occurrence (four of six for red and two of six for green). That is, predict twice as many reds as greens.

Strategy B: Predict the more likely color (red) on all of the 90 rolls.

Which strategy is best?

This test is based on one used by West and Stanovich (2003). Psychologist and economists have been giving tests similar to this in dozens (if not hundreds) of experiments since 1950. They find that many people (in some cases more than half) do not prefer or use Strategy B.

If you adopt Strategy B your prediction will be correct 2/3 of the time, your winnings will average 2/3 × \$90 = \$60. But if you adopt Strategy A, then you will be correct 2/3 of the time on average on your 60 picks of red. You will be correct 1/3 of the time on your 30 picks of green. Your average winnings will be 2/3 × \$60 + 1/3 × \$30 = \$50.

Strategy A yields an average of \$50. Strategy B yields an average of \$60. \$60 is greater than \$50, so Strategy B is best. The decision Strategy A is called "probability matching," since it involves matching the proportion of your decisions in favor of each option to the probability of that option.

Do not pick seed upsets according to the historical rates.

9.2 THE PERSISTENCE OF PROBABILITY MATCHING

Researchers have tried various methods to convince subjects to always use Strategy B, with limited success. A study of college students finds that those who have taken one or more college statistics courses pick Strategy B no more often than those that have had no college statistics courses (West and Stanovich 2003).

Probability matching is viewed as one of the more striking demonstrations of people's inability to reliably learn how to make rational financial decisions. As Nobel economist Ken Arrow put it: "The remarkable thing about this is that the asymptotic behavior of the individual, even after an indefinitely large amount of learning, is not the optimal behavior … We have here an experimental situation which is essentially of an economic nature in the sense of seeking to achieve a maximum of expected reward, and yet the individual does not in fact, at any point, even in a limit, reach the optimal behavior" (Arrow 1958).

Arrow's assessment may be a bit too pessimistic. A study by Shanks et al. (2002) found that approximately 71% of subjects learned to make reliable rational choices when they received extensive training. This included both structured feedback on their performance (compared to optimal performance) during the task and explicit nontrivial payoffs for each choice. But bracket pool play does not provide anything close to these conditions for learning. A bracket pool player can get feedback from noticing that their early round picks underperform the historical seed win rates, or the betting market predictions, or predictions based on a good rating system. But few players seek out this information. Bracket pool players don't get a direct payoff for each correct pick. The cause and effect relationship between picks and payoffs is probabilistic and obscure. Players get limited training as a result of playing the bracket pool. They only make 63 choices per year filling out one bracket, nowhere near the 1500 to 1800 choices that Shanks et al. used in their training exercises.

Looking at the poor past performance of your upset picks may help you break the habit of picking too many upsets.

9.3 THE PSYCHOLOGY OF FIRST-ROUND BRACKET PICKS

The problem of predicting how this two-colored die will fall resembles the problem of making your first-round picks in the bracket pool. Using historical seed performance to estimate the probabilities, a 5 seed wins about four of six and the 12 seed wins about two of six.

Psychologists Sean McCrea and Edward Hurt noted the similarity of typical bracket picking behavior to probability matching and decided to study the phenomenon in the laboratory (McCrea and Hurt 2009). The psychologists noted the suboptimal picks in the ESPN bracket contest. The strategy of simply picking the higher seed was 87.5% correct in the 2004 ESPN contest and 75% correct in 2005. The ESPN contest participants were only 75.2% correct in 2004 and 72.9% correct in 2005. The participants evidenced no superior ability to pick upsets. Their rate of correctly picking upsets was no better than chance.

McCrea and Hurt reported on three laboratory studies. The first was a study of NCAA bracket pool decision making, with a focus on decisions about first-round picks in each region. The two other studies involved decision-making tasks similar to making bracket picks for the eight first-round games played in one region of an NCAA tournament.

In the first study, they ran a laboratory bracket contest for 19 subjects. These subjects had a high interest in and knowledge of college basketball. The subjects were selected from a larger group of recruits based on their performance on a college basketball trivia test. The idea was to eliminate the possibility that the poor bracket picks evidenced in the ESPN contest were due to lack of basketball knowledge or lack of motivation in a large contest where there was little chance of winning. Each subject was paid $5 for participation in the study and the subject with the most accurate picks won $20. The psychologists focused on analyzing the first-round picks. The study participants picked few upsets of the 1 through 4 seeds. Their upset pick rate was similar to ESPN contest rate for the 5 through 8 seeds. Based on these results, probability matching in the ESPN contest did not seem to be due to lack of basketball knowledge or lack of motivation.

The second study was performed at a German university. There were eight outcomes designated using 16 ranks, 1 through 16. The outcomes were always designated 1 versus 16, 2 versus 15, 3 versus 14, 5 versus 12, 6 versus 11, 7 versus 10, 8 versus 9. But the source of the outcome was presented as different. In one study condition, they were presented as outcomes from basketball games of ranked teams, but the teams were only identified by their rank. No team names were provided. Only information on past performance of teams with the same rank was provided. The ranks, of course, correspond to seeds, but the term "seed" was not used to designate the ranks. In another study condition, the outcomes were presented as outcomes from random computer simulations, but the past performance information was exactly the same as for the basketball condition.

The second study had a number of goals. Using generic ranked teams eliminated some extraneous factors related to subjects liking teams or subjects having additional information or opinions about team strengths. The two conditions were designed to see if subjects used different prediction strategies for a human activity like basketball as opposed to outcomes randomly generated without a human performance factor.

Participants who thought that they were predicting basketball game outcomes used probability matching. The results for subjects predicting computer simulations did not follow the probability matching pattern. But the results for the computer simulation condition were surprising in another way: about 30% of the subjects picked the option ranked 16th instead the option ranked 1st even though the information provided on past performance indicated that the option ranked 16th had never been selected in 84 previous simulations. This seemed to indicate that some of the subjects thought the simulation outcomes were so random that they were not predictable from the previous results that were given.

In a third study, the computer simulations were replaced by a more concrete description of the process where paper slips were randomly sampled from envelopes as in a lottery. There were more slips with the better rank, so it was more likely to be picked. The participants were not told the number of slips of each rank, but they were told the results of 84 previous drawings. For instance, they were told that a slip of rank 16 had never been picked. Describing the process as a lottery did lower the randomness of the predictions. The predictions did not fit the probability matching pattern. The third study also included a condition where more information was provided about the unnamed basketball teams, including win/loss record, schedule difficulty, percentage of shots made, and percentage of opponent's shots made. Including more basketball information did not significantly change the predictions.

In addition to tabulating the subjects' picks, the researchers quizzed the subjects about their strategies. Subjects who thought they were predicting basketball games were more likely to report that they were trying to make correct picks. The subjects considered a basketball game to be more predictable than results from a computer simulation or a lottery. Based on the information provided to the subjects, the basketball games, the computer simulations, and the lotteries represented exactly the same probability models.

The researchers concluded that this was evidence that probability matching can arise from setting a goal of making 100% correct predictions. For instance, we can see from past tournaments that one or two of the 12 seeds score upsets in the first round of almost all NCAA tournaments. It follows from this fact that making four correct picks for next year's tournament will likely involve picking one or two 12 seeds to upset their first-round opponent.

Don't try to beat the averages by picking upsets in the early rounds in a standard scoring pool.

9.4 PROBABILITY MATCHING IN THE MEDIA

If you start searching the Internet for advice, you will find many well-respected sources that advise you to use a strategy equivalent to probability matching to make your decisions about first-round picks in bracket pools.

- *National Public Radio:* "If you want to pick upsets, 12 seeds for some reason do well against fives. So go ahead and pick them" (Goldman 2017).

- *CBS:* Chris Wittyngham recommended picking seven first-round upsets involving seeds in the 10 to 15 range because: "On average there are seven double-digit upsets per year" (Wittyngham 2017).

- *The NCAA joins the chorus:* "The world is not picking the 12 seed or 7 seed enough. No. 12 seeds win almost 36 percent of their games, but we pick them only about a quarter of the time. Seven seeds (seemingly one of the most enticing picks to be upset) win 61.4 percent of their games but are only picked to win 51.2 percent. That difference of 10.1 percent is the greatest of all first-round matchups … You can go with the higher seeds in the first round, but you probably want a total of six upsets among teams seeded 10–15. History shows that's about how many of those lower seeds actually win in the first round" (Benzie 2018).

9.5 PROBABILITY MATCHING IN ACADEMIA

BracketOdds (Bracketodds 2018), the website for a bracket analysis project at the University of Illinois, gives similar advice: "… an average of 1.68 teams seeded No. 13 or lower (worse) have won in this round, so picking one such upset is prudent." But elsewhere on the site, the advice is to pick no 15 seeds and no 16 seeds to win and pick one or no 13 seeds or 14 seeds. In the 10–12 seed range, the advice is: "Picking one, two or three makes sense, with a slight preference for two." They also recommend limiting the number of 1 seeds to advance to the Final Four: "Picking one or two No. 1 seeds in the Final Four is prudent." All this is based on what they call "patterns that have consistently emerged to help you avoid a bad bracket."

Computer science professor Sheldon H. Jacobson oversees the BracketOdds website. He intends for you to use only historical seed information. He said, "History suggests that you will be no worse—nor better—than trying to do it based on knowledge of the teams" (Touchstone 2017). No need to peek at the betting market spreads or team ratings to help you determine which team to pick according to Jacobson. But, Jacobson has also acknowledged that basketball rating systems have predictive value (Jacobson and King 2009).

The BracketOdds project grew out of a research project at Illinois. "What started (in 2007) as a simple inquiry on how seeds advance in the NCAA men's basketball tournament has grown into a Science, Technology, Engineering, and Mathematics (STEM) Learning Laboratory in Computer Science at the University of Illinois at Urbana-Champaign. Both undergraduate and graduate students have been involved in conducting research, analyzing data, and shaping this web site (first launched in 2011) into an informative tool for March Madness neophytes all the way up to skilled bracketologists." Bracket pool analysis at universities is typically used as a teaching or motivational tool. And the site does provide interesting information on historical patterns in the tournament outcomes.

But then BracketOdds starts to give advice based on these patterns. One problem is that more accurate data is available on the specific teams in each year's tournament. It's better to use data on the team than to just rely on generic data on teams with the same seed from past years. But their worst mistake is their nearly relentless use of probability matching as their strategy for making decisions about which seed to pick.

"There are patterns that exist in the seeds," Jacobson says. "As much as we like to believe otherwise, the fact of the matter is that we've uncovered a model that captures this pattern. As a result of that, in spite of what we emotionally feel about teams or who's going to win, the reality is that the numbers trump all of these things," Jacobson said. "It's more likely to be 1, 1, 2, 3 in the Final Four than four No. 1's …I don't know which two 1's, I don't know which No. 2 and I don't know which No. 3. But I can tell you that if you want to go purely with the odds, choose a Final Four with seeds 1, 1, 2, 3" (Ahlberg 2011).

But, the problem with this reasoning is that you don't get points for merely getting the pattern right. Your score depends on a specific permutation of this pattern. There are 12 permutations of 1, 1, 2, 3. Even if you get the pattern right and pick one of the 12 permutations at random, you will get, on average, only 1.5 picks correct even if the tournament outcome matches the pattern. See Table 9.1. One of the 12 permutations gets 4 picks correct, five get 2 picks correct, four get 1 pick correct, and two get zero picks correct. This averages out to 1.5 correct picks.

You get 1.5 picks right if you play the 1,1,2,3 pattern and the tournament outcome fits that pattern. But the tournament outcome will not always match that pattern. In general, if you play the 1, 1, 2, 3 pattern, you will average only 1.308 correct picks. You will get 1.940 picks correct on average if you pick all four 1 seeds. These averages are based on historical seed advancement rates (1985 through 2018) found at BracketOdds. The 1 seeds advanced to the Final Four 48.5% of the time. The 2 seeds advanced 19.1% of the time and the 3 seeds advanced 14.7% of the time, $0.485 \times 4 = 1.940$ and $0.485 + 0.485 + 0.191 + 0.147 = 1.308$.

Granted, there might be a situation where a specific permutation of 1,1,2,3 constitutes an ordered set of good contrarian picks for the Final Four, but that would be based on the win probability and opponent pick popularity of specific teams in specific years.

TABLE 9.1   The 12 Permutations of the Pattern 1,1,2,3 with Correct Picks
Assuming the Tournament Outcome for the Four Regions is East = 1, South = 1,
Midwest = 2, West = 3

Permutation Number	Permutation Random Pick Probability	East	South	Midwest	West	Correct Picks
1	1/12	1	1	2	3	4
2	1/12	1	1	3	2	2
3	1/12	1	2	1	3	2
4	1/12	1	2	3	1	1
5	1/12	1	3	1	2	1
6	1/12	1	3	2	1	2
7	1/12	2	1	1	3	2
8	1/12	2	1	3	1	1
9	1/12	2	3	1	1	0
10	1/12	3	1	1	2	1
11	1/12	3	1	2	1	2
12	1/12	3	2	1	1	0
Total						18
Average						18/12 = 1.5

Given the website's deficiencies, you might be tempted to recommend BracketOdds to your office bracket pool opponents on
the theory that using it will make them easier to beat. But, this
might backfire on you. BracketOdds also encourages making
the champ picks based on historical rates. This leads to much
less over-betting of the most popular 1 seeds and more contrarian champ picks. So, a bracket recommended by BracketOdds
may actually be an improvement over the typical opposing
bracket.

BracketOdds also provides analysis of the strategy that is the
direct opposite of seed-based probability matching. The page
"Why Pick Favorites" provides information on a strategy, dubbed
"Pick Favorite," that involves advancing all the better seeds. This
gives you a Final Four with all 1 seeds. There are only eight possible distinct brackets with all the 1 seeds advanced as deep as possible. The page gives the strategy a lukewarm endorsement: "If you

want to avoid being at the bottom of your bracket pool, use Pick Favorite to create your bracket." They also say: "In some years, it has produced a very good bracket." The page tabulates the scores of all possible Pick Favorites brackets for all years since the tournament expanded to 64 teams. The scores are based on the 2018 ESPN standard scoring system (10, 20, 40, 80, 160, 320). Dividing these scores by 10 provides the scores for the exponential scoring rule. If your office pool uses this most common scoring rule and you have access to the winning scores for past years, you may find that one of the eight Pick Favorites strategies would have finished in the money a number of times in your pool. According to the BracketOdds tabulation, the strategy of betting all eight Pick Favorites brackets would have finished in the money four times in the past 11 years (2008–2018) in the pool analyzed in Chapter 9.

9.6 HOW TO PICK FIRST-ROUND SEED UPSETS

There are only two good reasons for picking seed upsets in the first round of standard scoring pools. First, not all seed upsets are real upsets. If the betting market point spreads indicate that the lower seeded team is the stronger team, then this is typically a good reason to pick a seed upset in the first round.

The second reason is that there can be a minor advantage in picking an upset in the first round based on contrarian considerations. If the objective probability of an upset is at or very close to 50% and the underdog is a very unpopular pick, then there can be a slight advantage in picking the underdog. For instance, 12 seed Texas A&M and 11 seed George Mason were the best first-round picks for the 2006 ESPN's Tournament Challenge according to Clair and Letscher's parametric method (Clair and Letscher 2006) using their Sagarin-based tournament outcome model. These teams were both slight underdogs with win probabilities of approximately 48% and both were very unpopular picks with pick rates of 19.1% and 12.1% in the opponent pick distribution. But neither team was best pick for Yahoo's smaller contest. The Yahoo entry was based on an estimated 1,000,000 opponents whereas the

ESPN entry was based on 3,000,000 opponents. Perhaps the first-round picks carried more weight in the ESPN contest because the 2005 ESPN scoring rules (10, 20, 40, 80, 120, 160) were less top-heavy than the Yahoo scoring rules (1, 2, 4, 8, 16, 32).

The best policy is to base first-round picks on betting market spreads. If the betting markets indicate even odds, then pick the less popular team according to the Yahoo or ESPN pick distribution (adjusted for hometown bias). Otherwise, picking seed upsets seems to be unfavorable, with perhaps an uncertain grey area in the case of very slight underdogs that are very unpopular picks. But, this grey area, if it exists, will provide only a small advantage at best. Over all, this policy will result in picking few first-round seed upsets, and none of them will be based on probability matching or historical seed performance.

Picking early round upsets is rarely justified and never plays a large role in increasing your win chances in a standard scoring pool.

But, given the influence that choice anomalies like probability matching have over human behavior, it seems unlikely that we will see widespread optimization in first-round bracket choices in standard scoring pools. As Brad Carlin once told the *New York Times*: "The brain is one of the least effective predictive machines we have" (Leitch 2007).

CHAPTER 10

Bracket Advice Sources

T HERE SEEM TO HAVE been no sources of good bracket advice before Metrick's 1996 paper on standard scoring pools and Breiter and Carlin's 1997 paper on pools with upset incentives. The first book that Google indexes that mentions NCAA tournament bracket pools recommended picking game winners at random (Wayne 1984). It took a while for the findings from those first two research papers to filter out to the public in general. The typical bad advice had you betting too many early round upsets, with little or no emphasis on betting a mild upset for the overall champ pick.

Stephanie Williams wrote a good early article for *SmartMoney* magazine in 2001. She interviewed Brad Carlin, Ed Kaplan, and myself. She recommended using the Vegas odds to pick the first round and relying on power ratings for later rounds and picking a contrarian champ. This is all good advice for a standard scoring pool. For pools with upset incentives, she recommended the Kaplan and Garstka method that was available at Poologic.com. But she also mentioned some dicey approaches to picking more upsets to differentiate your bracket from the competition. David Leonhardt's 2005 *New York Times* article was the first non-research publication to completely avoid errors. Except for recommending a contrarian champ, Leonhardt refrained from offering advice on picking upsets in a standard scoring pool. This was before Clair and Letscher's 2007 research paper that shows that picking early round upsets without proper analysis may well make your bracket more like those of the competition. The notion of an all-chalk down-bracket was more of a theory or heuristic before Clair and Letscher provided some actual evidence on the optimal down-brackets in standard scoring pools. Websites that offer good bracket pool advice started becoming available after 1999.

These days, good advice coexists beside bad advice on the Internet. The bad advisors typically make poor use of team strength base rates, advise betting too many upsets in the early rounds, fail to use quantitative information at all to estimate opponent play, and recommend wrong-headed contrarian strategies.

10.1 GENERAL CHARACTERISTICS OF POOL ADVICE SITES

These sites will give similar advice for similar inputs. The inputs allow you to specify the characteristics of your office pool. The optimal bracket will be different depending on a number of characteristics of your pool: the number of opposing entries in the pool, the scoring rules of your pool, the local favorite teams, the number and relative size of the prizes for finishing first, second, third, and so on. If you submit multiple brackets in a pool, the optimal bracket will depend on the other brackets you are submitting.

All the bracket optimization tools have limits on the pool characteristics that they allow the user to specify. Team Rankings is the most full-featured for standard scoring pools. Poologic supports a few types of upset incentives that are not supported by Team Rankings.

If you use a free site, then your ability to specify pool characteristics will be limited. The tool is making a default assumption and not allowing you to change the default. If the tool does not allow you to specify the prize money breakdown, then it is probably optimizing the bracket on the assumption that one winner will take all of the pot. If the tool does not adjust for local favorites, then you should avoid betting a local favorite for champ. If you want to bet multiple entries and the tool does not support advice on multiple entries, then betting different strong champs should provide a reasonably diverse set of entries.

10.1.1 Estimated Return on Investment

Pool advice sites will typically claim that you can do two or three times better than the average pool player in a standard scoring pool. That is the estimated average over many years of pool play. This is a bit lower than the return on investment (ROI) estimated in some of the peer-reviewed research papers. Those estimated averages are based on the assumption that the tournament outcome model and the opponent models are perfectly accurate. The averages quoted by the pool advice sites are based on real-world experience and/or some hedging to ensure that they don't overclaim on the effectiveness of their bracket optimizers. The models are not perfect and that reduces the actual ROI relative to the theoretical ROI. Performance two to three times better than the average player is a reasonable expectation when submitting one optimized bracket in the typical office pool with standard scoring. And that assumes that others in your pool are not using state-of-the art analytics to optimize *their* brackets.

10.1.2 Simulation Algorithms

A number of the advice sites use Monte Carlo simulation methods. The specific methods are proprietary in most cases. Except for Poologic, none of the advice sites reference the research literature on bracket optimization. (mRchmadness references the research literature in their plans for future improvements.)

The core simulation method is likely similar to the Clair and Letscher framework but with the normal approximation of the ROI replaced by Monte Carlo estimation of the ROI. This method is based on simulating the distribution of tournament outcomes using a tournament outcome model, simulating your opponent's brackets using an opponent model and your estimate of the size of the pool, and then estimating the ROI of candidate brackets using these simulations along with information you provide about your pool such as the pool scoring rules. A hill-climbing algorithm is used to estimate the optimal bracket. Base on their results and their limited descriptions of their methods, it appears most sites that use simulations use some variation of this core method, perhaps along with some proprietary improvements.

10.2 BRACKET VOODOO

Bracketvoodoo.com provides a free bracket optimizer that you can use to optimize and evaluate brackets, and a more full-featured Voodoo Bracket Optimization Engine for paid subscribers. The required inputs are the size of your pool and the pool scoring rules. The tool supports standard scoring based on the exponential scoring rule, the Fibonacci scoring rule, and round number scoring (1, 2, 3, 4, 5, 6). The only upset incentive rule that the site supports is round number plus seed number.

The key output from the analyzer is the probability that your bracket will win your pool. The tool does not provide analysis that takes hometown bias into account, so you may want to adjust your bracket if a local team is promoted to the championship slot.

The free tool allows you to modify the optimized bracket or enter your own bracket from scratch and estimate the win probability of that bracket. This feature allows you to test some of your own preferred picks and see if you can come up with a personalized bracket that has a good probability of winning your pool. To assist you in coming up with picks that don't hurt your win probability too much, it puts a star on pivotal games and indicates dicey (i.e., ill-advised) picks with a red or orange border. "Pivotal games are the ones that can have the greatest impact on your win probability." It's informative and fun to play around with the free bracket optimizer.

The optimizer results are based on simulations. The Bracket Voodoo optimizer uses "advanced genetic algorithms and other optimization techniques." It is unclear what the utilized genetic algorithms might be. The nearest-neighbor hill climb originated by Clair and Letscher has the characteristics of a genetic algorithm in that the bracket evolves by "mutation" of a single team and survives in competition with its nearest-neighbor brackets if the mutation has the higher ROI.

The free content includes an informative and entertaining blog. The Bracket Voodoo free optimizer is one of the free full-bracket optimizers for standard scoring pools. It is a full-bracket optimizer in that it uses an opponent model to optimize all games in the bracket, not just the championship game. Poologic only optimizes the championship game pick based on an opponent model and uses the expected-point-maximizing teams to fill out an all-chalk down-bracket.

10.3 DAVID LETSCHER'S PICKS

This optimizer has not been available in recent years. For a number of years, David Letscher provided brackets for a range of standard scoring pool sizes and outcome models that were optimized using the Clair and Letscher (2007) method. You can, however, download the software for the method from http://math.slu.edu/~clair/pools/.

10.4 MRCHMADNESS

mRchmadness provides numerical tools for filling out an optimized bracket. It is an R package for use in the R statistical computing environment. R and mRchmadness are freely available under the GNU public license. The package authors are Eli Shayer and Scott Powers. The first version was made available in 2017. This bracket optimizer supports optimizing brackets for the men's and women's tournament. The optimizer finds the best bracket in a set of randomly selected candidate brackets rather than using a hill-climbing algorithm. You can use mRchmadness on the web at this site: https://saberpowers.shinyapps.io/mRchmadness/.

10.5 POOLOGIC

Poologic.com debuted in 2000 in time for the March bracket pools. Poologic is a free site that encourages users to donate some of their pool winning to the V Foundation for Cancer Research. This is the charitable foundation founded by Jim Valvano, the coach of the "Cardiac Pack," the 1983 NC State Wolfpack team that won an improbable run of close games to become the NCAA Division I champions. Poologic is not affiliated with the V Foundation.

Poologic provides an easy-to-use implementation of the algorithm from Kaplan and Garstka (2001) for finding the expected-point-maximizing bracket. It covers a wide range of pool rules for upset incentives. For standard scoring pools, Poologic provides a calculator that helps the user pick a contrarian champ. The calculations provide advice roughly similar to the advice in Metrick's 1996 paper, except that Poologic allows the user to assign a higher value to an all-chalk down-bracket than did Metrick. For a given pool size, Poologic is more prone than the Metrick algorithm to recommend an unpopular 1 seed rather than a 2 seed as the champ pick. It typically recommends against picking the most popular team for champ unless it is the strongest team and your pool is small.

10.6 SMARTBRACKET

SmartBracket (smartbracket.io) is a web application developed by Supported Intelligence, LLC. It is implemented using their Rapid Recursive© Toolkit. It's a bracket optimizer for standard scoring pools. The SmartBracket optimizer allows you to specify up to three teams that are popular in your pool. It also allows you to specify up to three teams that are overrated (in the opinion of the user) and three that are underrated. Access to the bracket optimizer requires a small fee.

The details of the method are not fully revealed, but the website provides some clues. It seems to involve a unique algorithm for standard scoring pools that is not based on simulations. SmartBracket maximizes the relative expected score of the bracket without regard to the size of your pool. The other bracket optimizers for standard scoring pools have the goal of optimizing your expected ROI or your expected win probability.

The SmartBracket documentation provides an example of how to calculate the relative expected score for a single game. Suppose a correct first-round pick is awarded 1 point in your pool. Suppose Kentucky and Florida are playing in the first round and Kentucky has a 90% chance of winning and a 100% chance of being picked by your opponents. If you pick Kentucky, then you will end up with the same number of points as all of your opponents regardless of whether or not Kentucky wins or loses. So, if you pick Kentucky, your relative expected score for that game will be zero.

Now consider the case where you pick Florida. If you are right, your relative score will be 1 versus your opponents. If you are wrong, your relative score will be −1. You will be right 10% of the time and wrong 90% of the time. So, your relative expected score will be $1 \times 0.1 + -1 \times 0.9 = -0.8$. SmartBracket uses the game-specific relative expected scores in a recursive algorithm to maximize the overall relative expected score of your bracket.

Kaplan and Garstka (2001) uses a recursive algorithm to maximize the absolute expected score for pools with upset incentives without regard to pool size. SmartBracket uses a recursive algorithm to maximize the relative expected score for pools without upset incentives without regard to pool size. On the face of it, it sounds like the SmartBracket folks may be applying the algorithm originally developed by Kaplan and Garstka to the relative expected score instead of the absolute expected score.

SmartBracket does not tailor its contrarian picks based on your pool's size. The website says that SmartBracket is designed to perform well in small local and office pools.

10.7 TEAM RANKINGS

Teamranking.com has been giving bracket advice since 2000. Access to specific bracket recommendations requires a fee. The publicly available bracket tips from past years indicates that Team Rankings was using an opponent model for standard scoring pools by 2009. They expanded their bracket pool research efforts in 2011 (Eden 2018).

Team Rankings uses data from the ESPN and Yahoo contests in their opponent model. They use ratings and betting market data to estimate team strength. The relative performance of candidate brackets is evaluated using simulated tournament results and simulated pools. Brackets are "mutated" and "mated" in an evolutionary process until no further improvement is possible (Eden 2018). This sounds similar to the Clair and Letscher nearest-neighbor hill climb, except that reproduction is asexual in Clair and Letcher's method, their simulated brackets don't mate.

The free content includes a blog that provides statistical information on bracket pools analyzed by the site's users among other topics. Approximately 50% of the pools analyzed have exponential scoring. The next most popular scoring rule is round number scoring (1, 2, 3, 4, 5, 6) weighing in at about 5%. Of pools analyzed, 19% have scoring rules that include upset incentives. The median pool size measures out at 60 entries. Pools with less than 40 entries accounted for 35.4% of pools analyzed (Barzilai 2015). In 2015, one Team Ranking customer reported entering 72 brackets across 12 different pools (Hess 2015).

The site supports a range of scoring rules that include upset incentives. Their recommendations are roughly consistent with the expected-point-maximizing algorithm developed by Kaplan and Garstka (2001), when tested for a 2018 pool with seed-multiplier scoring rules based on exponential factors in a pool with 60 entries and six payouts. The differences relative to Kaplan–Garstka could arise from the choice of the tournament outcome model or from toning down some of the riskier picks.

Team Rankings is currently the only site that optimizes brackets based on the payout structure of your pool. This can be an important improvement over a bracket that is optimized on the default assumption that it will be entered into a winner-take-all pool.

Team Rankings started as a Stanford University dorm room hobby of Mike Greenfield. The creators of Bracket Voodoo, mRchmadness, and the Power Rank are also Stanford graduates.

10.8 THE POWER RANK

Ed Feng developed thepowerrank.com after becoming interested in applying his PhD research to sports prediction. Feng uses a tournament outcome model based on a ranking system similar to the Google PageRank algorithm. During the period where you are determining your bracket, the site provides win probabilities for each team as part of its paid content.

Feng is the author of a book on how to improve your odds in a standard scoring bracket pool (Feng 2015). His book provides a walk-though of an expert approach to picking a good bracket. He recommends avoiding standard scoring pools with a large number of entries because you are too unlikely to win even with best play. He recommends just picking the strongest team as your champ if your pool is small. For pools in the Goldilocks middle, neither too big nor too small, he recommends finding a contrarian champ by considering all teams with a good chance of winning and finding one that is a relatively unpopular pick. He fleshes out the details of the method with numerical estimates based on simulations. The simulations use an opponent model based on ESPN data and his own tournament outcome model.

10.9 CONCLUSION

Is it worthwhile to pay for bracket pool advice? Team Rankings makes the argument that doubling or tripling the returns of the average pool player makes their fee worthwhile. But, if you are already making good use of the methods outlined in this book and the free information provided by advice sites, then you're play will already be above average. As part of their paid content, Voodoo Brackets and Team Rankings both provide recommendations for submitting multiple brackets that may be a cut above the advice available elsewhere. Team Rankings is the only site that provides simulation-based advice on pools with multiple payouts. Including multiple payouts in the analysis will tend to move you more toward betting more popular champs that are more likely to win. Some of the sites claim that their proprietary tournament outcome model is superior, but they don't provide statistical proof. It's hard to provide such proof of a marginal improvement. Both Team Rankings and Voodoo Brackets have looked into information beyond the pick distribution of nationwide bracket contests in an effort to validate and improve their opponent models, and some improvements may have resulted from this effort.

Paying for advice may provide some marginal improvements that may be worth the cost. You may find it more convenient to use a site with a paywall if you don't enjoy the effort of trying to work out the best bracket using the free information. And if you just want to have all the available information at your fingertips, then you will have to pay for some of it.

CHAPTER 11

Basketball Knowledge Considered Harmful

THE INVENTORS OF THE bracket pool wanted a game that would "determine once and for all who knew the most about college basketball," as Tim Trowbridge put it. But after a few years, the nationwide running joke was that the office bracket pool winner was that person in the office who knew the least about college basketball. Most people probably attribute this phenomenon solely to the uncertainty of the tournament outcome. But there is evidence supporting the existence of a cause and effect relationship between too much basketball knowledge and suboptimal bracket pool play.

Psychologist Tina Kiesler and her research team (Kiesler et al. 2001) found that there was an inverted-U curve relationship between performance on a 25-question basketball knowledge exam and performance on picking game winners in the NCAA tournament. This inverted-U curve, depicted conceptually in Figure 11.1, appears to mirror the frown of a basketball expert peering at it. The most knowledgeable subjects were actually the worst pickers of all. This team of researchers found the same

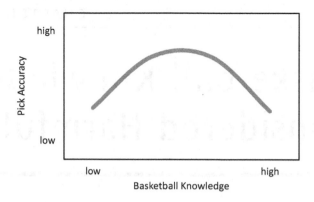

FIGURE 11.1 The inverted-U curve relationship between pick accuracy and basketball knowledge.

general pattern among subjects predicting outcomes in other sports, and they cited studies showing that experts perform no better than novices in some other domains (Morwitz 2018; Kiesler and Morwitz 2001).

How could this be? It turns out that prediction is a skill all its own, somewhat independent of the knowledge you possess on the matter being predicted. The best forecasters tend to give a lot of weight to *base rates* (i.e., past team performance). Kiesler et al. reported that "More knowledgeable subjects did not perform as well in this task because they too often relied on their own specific basketball knowledge and did not utilize base rates as often as they should." The researchers had evidence of this because they ask the subjects to provide reasons for their predictions.

Halberstadt and Levine (1999) tested the accuracy of self-identified basketball experts at predicting the outcome of the eight games played in the Sweet Sixteen round of the NCAA men's tournament for the years 1995 and 1996. The subjects were assigned to one of two groups. One group, the "reasoning group," was instructed to think about, analyze, and list the reasons for their predictions. The subjects were asked to list three reasons for each game. The other group, the "nonreasoning group," was instructed to use intuition or gut feelings, not think about the reasons for their predictions, and avoid drawn-out analysis. The pick accuracy of the nonreasoning group (70.4%) was significantly better than that of the reasoning group (65.2%). Both groups underperformed the pick accuracy of the betting market spreads (78.1%).

Haberstadt and Levine provided a couple of theories as to why analyzing the reasons for a pick might degrade accuracy. First, reasoning might bring more specific knowledge to mind, thereby making specific knowledge more likely to inappropriately override the base rates. Second, basketball experts might have better automatic prediction skills. The skill of picking the best team might be like riding a bicycle. If you are already relatively skilled at predicting basketball games, then too much thinking about it might make you worse.

The bracket pool is a game about a game. The basketball fans who invented the bracket pool overrated the importance of basketball knowledge in bracket pool play. And they did not realize the importance of specific strategies for the bracket pool itself independent of basketball. These strategies must be used to determine when it's favorable to pick the weaker team to win a tournament game. Basketball knowledge (knowing the win probability of the weaker team) is only one factor in deciding which upsets to pick. Estimating this probability and weighing other factors important in bracket pool strategy requires statistical reasoning. So, with almost 40 years of perspective on the bracket pool, it is perhaps more valid to say that the bracket pool determines not who knows the most about basketball but who knows the most about statistical reasoning.

References

Adams, T. 2017. Modeling Multiple Entry Strategies in the NCAA Tournament Bracket Pool. Poster presented at the New England Symposium on Statistics in Sports. http://nessis.org/nessis17/Adams.pdf (accessed May 31, 2017).

Ahlberg, L. 2011. Real March Madness is relying on seedings to determine Final Four. https://news.illinois.edu/view/6367/205387 (accessed October 4, 2018).

Allard, S. 2017. Is This Kent-Created NCAA Tourney Pool the Longest Continuously Running March Madness Contest in America? *The Cleveland Scene.* http://www.clevescene.com/cleveland/is-this-kent-created-ncaa-tourney-pool-the-longest-continuously-running-march-madness-contest-in-america/Content?oid=5741333 (accessed July 13, 2018).

AP. 1992. Cop Says No to Pool in School. *Seattle Times.* http://community.seattletimes.nwsource.com/archive/?date=19920329&slug=1483511 (accessed July 13, 2018).

Arrow, K. J. 1958. Utilities, Attitudes, Choices: A Review Note. *Econometrica* 26:1–23.

Barzilai A. 2015. The Most Popular (and Crazy) Types of Bracket Pools. https://www.teamrankings.com/blog/ncaa-basketball/bracket-pools-rules-scoring-systems (accessed June 15, 2018).

Bellhouse, D. R. 2011. *Abraham De Moivre: Setting the Stage for Classical Probability and Its Applications.* Boca Raton: CRC Press.

Benzie, M. 2018. March Madness: Can We Start Picking More 12 and 7 Seeds? The Data Says We Should. *NCAA.* https://www.ncaa.com/news/basketball-men/bracketiq/2018-03-14/march-madness-can-we-start-picking-more-12-and-7-seeds-data (accessed June 10, 2018).

Borghesi R. 2015. A Case Study in Sports Analytics: The Debate on Widespread Point Shaving. *Journal of Sports Analytics* 1:87–89.

BracketOdds. 2018. http://bracketodds.cs.illinois.edu/ (accessed June 10, 2018).

Bradshaw, Z. 2015. Predicting March Madness. http://blog.kaggle.com/2015/04/17/predicting-march-madness-1st-place-finisher-zach-bradshaw/ (accessed June 21, 2018).

Breiter, D. J., and B. P. Carlin. 1997. How to Play Office Pools If You Must. *Chance* 10:5–11.

Carlin, B. P. 1996. Improved NCAA Basketball Tournament Modeling Via Point Spread and Team Strength Information. *American Statistician*, 50(1):39–43.

Clair, B., and D. Letscher. 2005. 2005 Men's Basketball Tournament. http://math.slu.edu/~clair/pools/ncaa2005.pdf (accessed July 13, 2018).

Clair, B., and D. Letscher. 2006. 2006 Men's Basketball Tournament. http://math.slu.edu/~clair/pools/ncaa2006.pdf (accessed July 13, 2018).

Clair, B., and D. Letscher. 2007. Optimal Strategies for Sports Betting Pools, *Operations Research* 55:1163–1177.

Cohen, B. 2015. NCAA Tournament: Why You Should Pick against Kentucky. *Wall Street Journal.* https://www.wsj.com/articles/ncaa-tournament-whatever-you-do-dont-pick-kentucky-1426521564 (accessed October 15, 2017).

Daily News. 2007. Bracket-Bustin' Probes Shut S.I. Pub's Pool. http://www.nydailynews.com/news/crime/bracket-bustin-probes-shut-s-pub-pool-article-1.213977 (accessed July 23, 2018).

Edelman, M. 2017. The Legal Risk of Operating NCAA Tournament Pools. *Forbes.* https://www.forbes.com/sites/marcedelman/2017/03/13/the-legal-risk-of-operating-ncaa-tournament-pools/#6c25373f75b3 (accessed July 23, 2018).

Eden, S. 2018. The Key to the Perfect March Madness Bracket: Evolution. *Wired.* https://www.wired.com/story/the-key-to-the-perfect-march-madness-bracket-evolution/ (accessed July 2, 2018).

ESPN, eds. 2009. *ESPN College Basketball Encyclopedia.* New York: ESPN Books.

ESPN. 2016. BPI and Strength of Record: Reshaping the Debate of Best and Most Deserving. http://www.espn.com/blog/statsinfo/post/_/id/125983/bpi-and-strength-of-record-reshaping-the-debate-of-best-and-most-deserving (accessed October 10, 2017).

ESPN. 2018. Who Picked Whom. http://games.espn.com/tournament-challenge-bracket/2018/en/whopickedwhom (accessed April 13, 2018).

Feng, Ed. 2016. *How to Win Your NCAA Tournament Pool*. Self-published, Amazon Digital Services.

Fox News. 2009. Duke Coach to Obama: Worry About the Economy, Not NCAA Picks. http://www.foxnews.com/politics/2009/03/19/duke-coach-obama-worry-economy-ncaa-picks.html (accessed October 10, 2017).

Goldman, T. 2017. March Maddness 101: A Few Tips for Your Bracket. NPR. https://www.npr.org/2017/03/11/519845898/march-madness-101-a-few-tips-for-your-bracket (assessed June 10, 2018).

Halberstadt, J. B. and G. M. Levine. 1999. Effects of Reasons Analysis on the Accuracy of Predicting Basketball Games. *Journal of Applied Social Psychology* 29:517–530.

Hess D. 2015. The Results Are In: How Our 2015 Bracket Picks Did. https://www.teamrankings.com/blog/ncaa-basketball/how-2015-bracket-picks-did (accessed July 3, 2018).

Hill, B. 1997. Did Man Lead Way on NCAA Pools? *Courier Journal.* Reprinted in https://redragetailgate.com/2017/03/15/louisville-fan-invented-modern-day-march-madness-office-pool/ (accessed July 19, 2018).

Jacobson, S. H. and D. M. King. 2009. Seeding in the NCAA Men's Basketball Tournament: When Is a Higher Seed Better? *Journal of Gambling Business and Economics*, 3:63–87.

Jonietz E. 2002. Computing Athletics. *MIT Technology Review.* https://www.technologyreview.com/s/401706/computing-athletics/ (accessed September 16, 2017).

Kaplan E. H., and S. J. Garstka. 2001. March Madness and the Office Pool. *Manage Science* 47:369–382.

Kiesler T. and V. G. Morwitz. 2001. "Special Session Summary What Are the Chances? Biases in the Assessment of Probability and Risk." In Andrea Groeppel-Klien and Frank-Rudolf Esch, eds., *E - European Advances in Consumer Research Volume 5,* Provo, UT: Association for Consumer Research, 5:195.

Kiesler, T., V. G. Morwitz, and E.A. Yorkston. 2001. The Ball Bounces Differently for Experts and Novices. Paper presented at the Association for Consumer Research European Conference (June 2001).

Leitch, W. 2007. For gamblers, 1+2=5. *New York Times.* https://www.nytimes.com/2007/03/04/sports/playmagazine/04play-pool.html (accessed June 14, 2018).

Leonhardt, D. 2005. To Win the Pool, Look the Other Way. *New York Times.* http://query.nytimes.com/gst/fullpage.html?res=9F0CE6DD153CF930A25750C0A9639C8B63 (accessed October 15, 2017).

Mather, V. 2017. No N.C.A.A. Bracket for Trump, But How Good, Really, Were Obama's? *New York Times.* https://www.nytimes.com/2017/02/16/sports/ncaabasketball/ncaa-basketball-tournament-trump-obama.html (accessed October 3, 2018).

Mather, V. 2003. Personal Communication via Email.

Mather, V. 2011. How to Win a Pool That Rewards Upsets. *New York Times* http://www.nytimes.com/2011/03/14/sports/ncaabasketball/14pool.html (accessed November 10, 2017).

Matthews G. and M. Lopez. 2014. Q&A with Gregory Matthews and Michael Lopez, 1st Place in March ML Mania. http://blog.kaggle.com/2014/04/21/qa-with-gregory-and-michael-1st-place-in-march-ml-mania/ (accessed October 15, 2017).

McCrea, S. M. and Hirt, E. R., 2009. Match madness: Probability matching in prediction of the NCAA basketball tournament. *Journal of Applied Social Psychology* 39:2809–2839.

Medcalf, M. 2018. NCAA Tournament Selection Committee Testing New Tool That Could Eventually Replace RPI. *Wall Street Journal.* http://www.espn.com/mens-college-basketball/story/_/id/22144913/ncaa-says-year-test-run-new-evaluation-system (accessed June 22, 2018).

Metrick, A. 1996. March Madness? Strategic Behavior in NCAA Basketball Tournament Betting Pools. *Journal of Economic Behavior & Organization* 30:159–172.

Morwitz, V. 2018. Personal Communication via Email.

NCAA. 2010. Gambling on College Sports. http://www.ncaa.org/sites/default/files/bbd-gambling.pdf (accessed July 13, 2018).

NCAA. 2016. 2016-17 NCAA Division I Men's Basketball Championship Principles and Procedures for Establishing the Bracket. http://www.ncaa.com/content/di-principles-and-procedures-selection/ (accessed December 17, 2017).

Newmark C. 2003. Newmark's Door. http://www.newmarksdoor.com/mainblog/2003/03/my_wifes_workpl.html (accessed July 13, 2018).

Niemi, J. B. 2005. *Identifying and Evaluating Contrarian Strategies for NCAA Tournament Pools.* Master's Thesis. University of Minnesota, Minneapolis, MN.

Niemi, J. B., B. P. Carlin, and J. M. Alexander. 2008. Contrarian Strategies for NCAA Tournament Pools: A Cure for March Madness? *Chance* 21:39–46.

Null, B. 2016. Homer Bias Is Real and It Will Derail Your March Madness Bracket. https://www.cbssports.com/college-basketball/news/

homer-bias-is-real-and-it-will-derail-your-march-madness-bracket/ (accessed March 3, 2018).

Oliver, D. 2012. Introducing the BPI. *ESPN.* http://www.espn.com/mens-college-basketball/story/_/id/7561413/bpi-college-basketball-power-index-explained (accessed September 14, 2017).

Packard, E. 2001. Pakard's March Madness Contest. https://web.archive.org/web/20030814011035/http://home.mesastate.edu/~epackard/hoop/contest.html (accessed October 19, 2017).

Pauga, K. 2018. Comparing Advanced Metrics to NCAA 1-68 Seed List. http://www.kpisports.net/2018/03/13/comparing-advanced-metrics-to-ncaa-1-68-seed-list-2018-edition/ (accessed May 24, 2018).

Polya, G. 1973. *How to Solve It: A New Aspect of Mathematical Method.* Princeton: Princeton University Press.

Pomeroy, K. 2018 2018 NCAA Tournament Probabilities. https://kenpom.com/blog/2018-ncaa-tournament-probabilities/ (accessed May 24, 2018).

Rushin S. 2009, The Bracket Racket. In *ESPN College Basketball Encyclopedia*, ed. ESPN, 24–27. New York: ESPN Books.

Sagarin, J. 2002. Jeff Sagarin Ratings. *USA Today.* https://www.usatoday.com/sports/ncaab/sagarin/2002/teams/ (accessed September 16, 2017).

Sagarin, J. 2017. Jeff Sagarin Ratings. *USA Today.* https://www.usatoday.com/sports/ncaab/sagarin/ (accessed September 21, 2017).

Salsburg, D. S. 2017. *Errors, Blunders, and Lies: How to Tell the Difference.* Boca Raton: CRC Press.

Silver, N. 2012. *The Signal and the Noise: Why So Many Predictions Fail—But Some Don't.* New York: Penguin Press.

Shanks, D. R., R. J. Tunney and J. D. McCarthy. 2002. A Re-examination of Probability Matching and Rational Choice. *Journal of Behavioral Decision Making* 15:233–250.

Shayer, E. and S. Powers. 2017. *mRchmadness: Numerical Tools for Filling Out an NCAA Basketball Tournament Bracket.* https://cran.r-project.org/web/packages/mRchmadness/index.html (accessed March 5, 2018).

Sonas, J. 2015. How Did These Teams Get to the Top of the Leaderboard? https://www.kaggle.com/c/march-machine-learning-mania-2015/discussion/13007 (accessed August 13, 2018).

Stern, H. 1991. On the Probability of Winning a Football Game. *American Statistician* 45:179–183.

Stern, H. S. and B. Mock. 1998. A Statistician Reads the Sports Pages: College Basketball Upsets: Will a 16-Seed Ever Beat a 1-Seed? *Chance* 11:26–31.

Stigler, S. M., 1981. Gauss and the Invention of Least Squares. *The Annals of Statistics*, 465–474.

Tracy, M. 2017. Think the N.C.A.A. Bracket's Too Easy? Try One of These Pools. *New York Times*. https://www.nytimes.com/2017/0 3/13/sports/ncaabasketball/march-madness-pools.html (assessed August 13, 2018).

Thaler, R. H. 1993. *Advances in Behavioral Finance.* Russel Sage Foundation.

Thorp, E. 1961. A Favorable Strategy for Twenty-One. *Proceedings of the National Academy of Sciences* 47:110–112.

Thorp, E. 1966. *Beat the Dealer.* Random House.

Touchstone, L. 2017. Can Data Analytics Help You Fill Out a March Madness Bracket? https://blogs.illinois.edu/view/6367/471688 (accessed June 20, 2018).

Vander Voort, E. 2018. What Is a Team Sheet? Inside the March Madness Selection Tool. *NCAA.* https://www.ncaa.com/news/ basketball-men/bracketiq/2018-02-14/what-team-sheet-inside-march-madness-selection-tool (accessed June 25, 2017).

Wahl, G. 2005a. Pool Pointers. *Sports Illustrated.* http://web.archive.org/ web/20050316112909/http://sportsillustrated.cnn.com/2005/write rs/grant_wahl/03/15/bracket.picks/1.html (accessed November 10, 2017).

Wahl, G. 2005b. Personal Communication via Email.

Wall, A. 2015. All the President's Picks: See His 2015 NCAA Tournament Brackets. *White House.* https://obamawhitehouse.archives.gov/blo g/2015/03/17/all-presidents-picks-see-his-2015-ncaa-tourname nt-brackets (accessed July 13, 2018).

Wayne, J. 1984. *The Original Guide to Office Pools.* Toronto: Doubleday.

West, R. F. and K. E. Stanovich. 2003. Is Probability Matching Smart? Associations Between Probabilistic Choices and Cognitive Ability. *Memory & Cognition* 31:243–251.

Wilson, C. 2018. Pick Cincinnati and Bet against Virginia and Kansas in Your NCAA Pool. *Slate.* https://slate.com/culture/2018/03/pick-c incinnati-and-bet-against-virginia-and-kansas-in-your-ncaa-p ool.html (accessed May 24, 2018).

Williams, S. 2001. Everybody in the Pool. *SmartMoney.* March 2001.

Wittyngham, C. 2017. 2017 NCAA Tournament Brackets: Using the Last Ten Years to Predict Upsets. *CBS*. http://www.cbssports.com/college-basketball/news/2017-ncaa-tournament-bracket-using-the-last-10-years-to-predict-upsets/ (accessed June 10, 2018).

Wolfers, J. 2006. Point Shaving: Corruption in NCAA Basketball. *American Economic Review* 96:279–283.

Wolff. A. 2003. A Madness to the Method. *Sports Illustrated*. March 2003.

Wolff. A. 2016. Six Things We Learned in Eight Years of Obama's Brackets. *Sports Illustrated*. https://www.si.com/college-basketball/2016/03/16/barack-obama-ncaa-tournament-bracket-march-madness (accessed August 13, 2018).

Yahoo. 2018. Pick Distribution. https://tournament.fantasysports.yahoo.com/t1/pickdistribution (accessed April 13, 2018).

Index

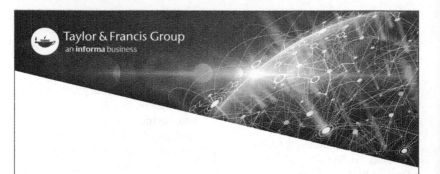